A MATTER OF
LIFE AND
DEATH

SUE ARMSTRONG

CANONGATE
Edinburgh · London · New York · Melbourne

Published by Canongate Books in 2010

1

Copyright © Sue Armstrong, 2010

The moral right of the author has been asserted

First published in Great Britain in 2008 by Dundee University Press

This revised edition published in Great Britain in 2010 by
Canongate Books Ltd, 14 High Street, Edinburgh EH1 1TE

www.meetatthegate.com

British Library Cataloguing-in-Publication Data
A catalogue record for this book is available on
request from the British Library

ISBN 978 1 84767 581 1

Designed and typeset in Sabon and Franklin Gothic by
Cluny Sheeler, Edinburgh

Printed and bound in Great Britain by CPI Mackays, Chatham ME5 8TD

'I really became interested in pathology during my second year at medical school. Looking through the microscope I saw medicine come alive. I wanted to understand disease and what caused it, and how these changes occurred in the patient. Pathology certainly was not a "dead" science in any sense. I felt I learnt most of my medicine by studying pathology. It was the basis for everything.'

Elaine Jaffe, Director, Laboratory of Pathology, Cancer Research Center at the National Cancer Institute, USA

CONTENTS

INTRODUCTION

I grew up in a medical family. At the time of my birth my father, a doctor of tropical medicine, was working in the southern Sudan and this is where I spent the first six years of my life. When independence came in 1956, my father left the Sudan Medical Service to join the Shell Oil Company. He moved to Borneo to run a specialist TB clinic and a general hospital serving the indigenous Dyak tribespeople, who lived in stilted long-houses on the fringes of big rivers deep in the jungles of Sarawak.

My father's was not a nine-to-five job. We lived close to where he worked; he was home frequently during the day, and my mother, two sisters and I were deeply involved in his world. I remember hearing his stories about patients and diseases at the meal table, accompanying him on ward rounds occasionally and being taken to meet particularly interesting people. There is a picture in the family album, for instance, of myself and my younger sister Julie standing at the hospital bedside of the Queen of the Dyaks. She is tiny and bird-like, and her pierced ears, weighted down with many rings, hang down to her waist.

My father was not a pathologist. But in the out-of-the-way places he worked, you had to be a Jack-of-all-trades; and he relied heavily on the voluminous pathology textbooks that lined his study walls. I have abiding memories of leafing through these with a mixture of horror and fascination at the lurid photographs of tropical ailments that disfigured faces and limbs. I remember, too, a photograph among the family collection from our Sudan days of a man with elephantiasis of the scrotum, who carried his massively swollen genitals before him in a wheelbarrow.

I might have followed in my father's footsteps but for two things: I am squeamish about other people's pain, and I always wanted to be a writer. But as a journalist it seemed the most natural thing in the world to focus on health and medicine. And when the opportunity came to explore this particular specialisation in depth, I was delighted.

Pathology – the scientific study of disease and disease processes – has its roots in Renaissance Italy, when some doctors, curious to know what lay behind the superficial signs and symptoms of illnesses for which they treated their patients, began to cut open the bodies of those who died, to explore the internal organs. But it was only around 200 years ago, under the influence of German physician Rudolf Virchow, that pathology developed into a formal discipline and began to be taught as part of the medical school curriculum.

Pathology is the cornerstone of modern medicine. It is the science that has progressively replaced the myth, magic and superstition of traditional medicine with a rational basis for the care of the sick. Pathologists are vital members of the clinical team, responsible for around 70% of all diagnoses in the UK National Health Service today. It is they who determine exactly what kind of cancer a patient is suffering from and what stage the tumour has reached; they who are responsible for recognising new diseases such as AIDS, SARS, bird flu and new variant CJD when they first appear in a population; they who identify the bacterium, virus or other organism responsible for an outbreak of infection in a community; and they who will be tasked with investigating why a seemingly healthy baby dies soon after birth.

The great majority of their work is to do with living patients, yet the prevailing image of the pathologist – popularised by TV – is of someone working with a body pulled from a canal or a shallow grave to try to find out what happened. It's this association with the macabre that led to their being vilified and dubbed 'doctors of death' when the storm blew up over practices at Liverpool's Alder Hey and Bristol Children's Hospitals in 2001. Then, parents of children who had died at the hospitals became aware for the first time that the small bodies they had received for burial after post-mortem were not always complete – that their children's brains and other organs were sometimes retained for further investigation or research. This was not a secret, but neither was it always made explicit to parents, and the revelation caused a storm of outrage directed at pathologists. Paediatric pathologists, in particular, came in for abuse. Some received threatening phone calls at home; others found their children being bullied at school because of their parents' profession.

To a medical journalist this was a challenge: how to defuse this ugly mood and bring some understanding of the critical, but hidden, role pathologists play in all our lives. I put the idea to the BBC of a radio feature looking at what kinds of people go into pathology and what they actually do. It was readily accepted and I was asked for a two-part series for Radio 4 that was broadcast at prime-time in the evening science slot. The first programme allowed pathologists simply to talk about themselves and their work. The second focused on a patient with breast cancer, examining the role played by pathologists in the management of her case, and following them into the lab where they were studying breast tissue from surgery. Making the programmes was fascinating: this is a profession that demands engagement with some of the most profound issues of living and dying – issues that fascinate us all, even if most of us choose not to dwell on them – and I came away from interviews thoroughly stimulated and inspired. Here was a story worth telling.

This book is a continuation of that story, and it takes a similar approach to the radio broadcasts – conversations with pathologists that explore the homes they grew up in, when and how their interest in pathology was kindled, who have been their mentors, heroes, role models, what their working lives entail, how their work on the front line of disease and death has affected their philosophy of life and how they view the prospect of their own demise. In pursuit of stories, I travelled north and south of the British Isles, zigzagged across the United States, flew to the southern tip of Africa and nipped across to Italy on a lovely spring day, to talk to pathologists in a wide range of settings and with a great diversity of personal experiences.

In Edinburgh, neuropathologist James Ironside told me of the first time he saw evidence that Britain's epidemic of mad cow disease was affecting humans. And he showed me under the microscope the difference between a so-called spongiform encephalopathy and Alzheimer's disease, though both conditions leave holes in the brain.

In London, paediatric pathologist Irene Scheimberg took me round the tangly garden of her terraced house before cooking me a Tunisian dish for lunch and telling me her life story – how she had had to flee from her native Argentina when the military junta started abducting her friends, and how she ended up doing pathology in London. Julia Polak, also from Argentina originally, talked to me in her apartment overlooking the Thames at Chelsea as seagulls screeched outside and the sound of boats chugging past drifted through the open French windows. Dame Julia, one of the longest survivors of a heart-lung transplant, was housebound that day, preparing for her biannual hospital check-up, and

4

she talked of her brush with death and her quest today to 'build' new organs for transplantation. In Oxford, paediatric neuropathologist Waney Squier shared her concerns about shaken baby syndrome and described the often theatrical nature of court hearings at which she has to bear expert witness over the death of a baby.

In Cape Town, South Africa, the only deaths that are investigated more or less routinely with autopsies today are those of babies, said Helen Wainwright. Based at the famous Groote Schuur Hospital, Dr Wainwright told me that two of the biggest scourges in the community she serves in the Western Cape are AIDS and alcohol, both linked to poverty and disadvantage. In the apartheid years, workers on the wine estates received part of their wages in alcohol – the so-called 'tot system', whose legacy is painfully apparent today.

I began my travels in the USA in Chicago, in a high-rise apartment block on the shores of Lake Michigan where I was to talk to Francisco González-Crussí. This was a man I was particularly keen to meet, having read his delightful essays on the pathologist's world in preparation for my radio programmes. Sipping the most delicate jasmine tea I have ever tasted – served to us in tiny, steaming cups by Dr González-Crussí's Chinese–American wife and fellow pathologist, Wei Hsueh – we talked of the often rocky road that led him from a very poor neighbourhood in Mexico City to a pathology professorship at Northwestern University Medical School and head of laboratories at Children's Memorial Hospital, Chicago.

In Boston I learnt of a lifetime's preoccupation with the fascinating conundrums of cancer from Christopher Fletcher, in his small office crammed with books and papers. And I met Kumarasen Cooper, a South African of Indian descent who told of the dreadful struggle to train as a doctor back home under the apartheid regime; of his own family's involvement in political opposition to the Nationalist government; and of his visits to his brother, incarcerated on Robben Island in a cell next to Nelson Mandela. In Washington I met Jeffery Taubenberger, head of a gleaming new high-containment laboratory for investigating lethal germs at the National Institutes of Health, whose personal research obsession has been the virus that caused the deadly Spanish flu pandemic of 1918.

After a few more stops and a lot more talking, I ended my travels in the US with a visit to the Body Farm in Knoxville, Tennessee. I had read about this extraordinary facility, where corpses are left to rot under varying conditions so that scientists can study the processes of decomposition, and could not resist the temptation to meet the man who started it in the 1970s. Bill Bass met me at Knoxville airport and, though I had expected to speak to him at his office or home, I found myself driven straight to the Body Farm for a conducted tour, followed by a visit to the archives where all the bodies finally end up as beautifully cleaned, preserved and catalogued boxes of bones. To my surprise, having been in the presence of death only rarely, I was not in the least disturbed by the experience. In that patch of woodland overlooking the Tennessee River, where the autumn sun shone dappled light on to bodies and bones among the fallen leaves and birds hopped about in the treetops, there was an air of peacefulness, timelessness and, yes, deep respect for the dead – for the people they had been and for the contribution they were making voluntarily to science. Bill Bass, now an old man, was gentle, courteous and great fun to be with as he shared his life story.

These are just a few of the fascinating characters I met. Without exception, my interviewees are people who are passionate about what they do: no one admitted to ever wishing he or she had chosen a different path in life.

However, only two of my interviewees had a clear intention to go into pathology before they entered medical school; the great majority had little idea of what a pathologist did and no great pull in that direction until they chanced upon an inspiring teacher, or became transfixed by the world that unfolded under the microscope. Very many described how singularly beautiful that world is, and, by extension, how visual appreciation and visual memory are essential to success as a pathologist.

This is one of several common themes to emerge from the conversations. Another is the intense curiosity felt by my interviewees as they trained to become doctors, to know not only what ails a patient, but how and why – to uncover the mechanisms of disease behind the diagnoses. But the popular image of the pathologist as a somewhat socially dysfunctional loner turns out to be a myth. While some people did say they were happier working behind the scenes than on the front line of clinical care, a surprising number do have direct contact with patients and families.

As in every field of science and medicine, the ethical standards that underpin pathology practice are not cast in iron, but evolve in line with the ceaseless debate in societies about what is right and wrong. Thus many of the things that were acceptable 10, 20, 30 years ago are not acceptable today. As these interviews show, context, too, is important: things that are acceptable in one culture may not be so in another. Mostly, the rules change gradually and piecemeal. But sometimes a dramatic event precipitates a wholesale rethink. The Alder Hey and Bristol controversy, which drew in a wider circle of institutions, was one such event. In the UK it changed the relationship between the people and their health service in a few short months, and may well have pushed the pendulum of reform too far and too fast. A major theme to emerge from the interviews is mighty frustration among pathologists with the tangle of red tape that surrounds pathology practice almost everywhere today, and the threat this represents to existing archives and future collections of samples and specimens of human tissue that are so vital to teaching, research and advances in medicine.

This book is essentially a collection of self-portraits, and though they share some common concerns, the people whose voices you will hear in these pages, and the stories they have chosen to tell about themselves, are all very different. For some, the emphasis is on the eventful journey to where they are now; for others, it is on the disease or research topic that has become the focus of their working lives.

But why, you might wonder, should a book that deals with the 'science of medicine' appeal to a general audience? When I was working on the radio programmes, I passed one day, on my way to an interview in Central London, the venue of the *Body Worlds* exhibition – the controversial show put on by German anatomist Gunther von Hagens, which gave visitors a tour of the human body preserved in lifelike manner by 'plastination' and presented in personal and sometimes shocking ways. A handful of angry protesters wielding placards outside had failed to deter attendance, and visitors were queuing round the block. Bodies – how they work and what can go wrong with them – are quite simply fascinating to us all.

Finally, a word about selection and editing. Only about half of the interviews undertaken for this project are featured here, simply because the conversations proved too interesting to curtail, and there was a limit to the length to which this book could run. With the subjects' collaboration, the interviews here have been significantly cut down, and the text sometimes reordered, to tell a much shorter story than the original, or else to focus on a particular aspect of the whole. But what has eased the painful task of selection somewhat is the fact that the full, rich collection of conversations, each running to its original length, is to be made available in an archive for the Pathological Society. Thus readers who wish to know more about the individuals and their lives, and/or the science with which they have engaged, will be able to access their interviews on the Internet (at www.pathsoc.org) soon after the book is released.

'PHYSICIAN, HEAL THYSELF'

Julia Polak
Director of the Tissue Engineering and Regenerative
Medicine Centre, Imperial College London

Julia Polak has been at the cutting edge of medical research since the late 1960s and is one of the most cited scientists in the world. She was among the first to demonstrate the existence of a hormone system in the gut and was on the team that discovered how nitric oxide – a substance produced by exhaust fumes and the magic molecule in Viagra – is made in cells throughout our bodies and governs a wealth of vital biological processes. In 1995 she was doing research into a rare lung condition, when she herself collapsed with the disease and had to have a heart-lung transplant to save her life. Later she presented her own case at a medical meeting at Hammersmith Hospital, and says of her lungs, 'They were lovely in terms of pathology . . . [But] it was an extreme case of pulmonary hypertension, and I couldn't imagine how I had managed to breathe!'

Dame Julia is one of the longest survivors of a heart-lung transplant in the UK, and the experience changed her life course. She switched to the newly emerging science of tissue engineering and regenerative medicine, which aims to find new ways of repairing damaged organs using patients' own stem cells, and will eventually replace the need for donor transplants.

Dame Julia grew up in Argentina, to where her Jewish grandparents had emigrated from Eastern Europe to escape persecution. She and her husband Daniel Catovsky, a haematologist, came to the UK in 1967 for a year's postgraduate study and never went back. The science in Britain was exciting and back home their own families were being persecuted by the military junta.

———•——

What was your first experience of seeing a dead body?

It was more or less when I started pathology. As a student I had to clean the pieces for the museum, and there were bits in formalin that I had to take out and in. I remember I had to come in very early one morning to the post-mortem room – it was a very cold morning, and when I reached into the bucket a hand clasped my hand. [*laughs*]

And how did you react?

With disgust! I ran out of the room.

But you psyched yourself up and went back in?

It wasn't really a crisis. I'm not that kind of person. I remember after my transplant, people said: 'You'll need psychotherapy, some support.' But I'm not the reflective sort.

When you chose pathology, was it the disease mechanisms that most fascinated you or were you focused on the patient and the patient's family?

The *why* of diseases is what interested me. I was interested more in understanding the mechanisms of disease and the science than curing patients. I mean, to such a degree that when they took my diseased lungs for research purposes at the Hammersmith, people asked, 'How *could* you look at your diseased lungs?' But you don't think 'That was part of me,' you look at the science behind it.

So tell me about your operation. How did you come to realise you needed a transplant?

It's a long story . . . I'd probably had the disease all my life, but nobody realised. I was becoming more and more breathless and people said, 'It's asthma.' But I had a very rare disease called pulmonary hypertension – high blood pressure in the lungs – where the incidence is one in a million. It was only diagnosed when I collapsed at the Hammersmith. I was in heart failure and it was terminal. But until then, no one had considered it.

How long had you been experiencing breathlessness?

All my life. I had probably had pulmonary hypertension all the time, although it was probably a very mild form until it suddenly deteriorated. I had my three children, and I was flying everywhere, giving lectures all over the world . . . But it wasn't comfortable, and at the end I couldn't even walk.

And did you ever suspect that's what it might be?

I didn't suspect pulmonary hypertension. I thought it was something more than asthma, but I couldn't tell.

But you'd been studying pulmonary hypertension, hadn't you? Why did it not occur to you?

I suppose I thought, 'It doesn't happen to me.' It could be denial; I can't tell. I'm not a very good psychologist.

So what exactly happened?

Well, I collapsed. That was in April 1995, and my transplant was on 17 July. I was 56 years old. When they got me to hospital they were shocked, obviously, because they went out of my room to talk, and I suppose they were discussing how to break the news. I couldn't understand why they'd gone outside. And then when they said, 'You have pulmonary hypertension,' I demanded: 'But *why*?' I was so angry I wanted to hit someone! They said, 'We don't know why. But we have to move you to the ward.' I said, 'No, no, I'm going home. Get my clothes.' They said, 'No, you're staying.' And then they called my husband Danny. He was shocked and didn't understand at first. And then they told him, 'She's seriously ill.' They put me on a drip and oxygen and all that. Then an expert from the Hammersmith, Celia Oakley, made a diagnosis. That was April, early May. They started treating me with something, but it didn't do any good; I was terminal, so I was deteriorating fast.

Were you in hospital all the time?

I was in hospital only for two weeks at first. They sent me home, but I collapsed there again so I went back, and then at a certain point Sir Magdi Yacoub, the famous transplant surgeon, appeared. I was carrying out research projects with him at the time, and I said, 'Why are you here?' He said, 'It's a social visit.' [*laughs*]

So they were keeping you in the dark?

Actually, they were in the dark themselves; they just couldn't believe what they were seeing.

And they thought you weren't going to survive?

No. But when they put the possibility of a transplant to us, we didn't want to accept it. Remember, this was 13 years ago; it's not the same as today. They didn't do so well in the early days, heart and lung transplants. I was in the bed, Danny was by my bedside, and one of my colleagues at the Hammersmith walked in. He had nothing to do with my case, but he knew we were resisting making a decision, and he said, in front of me, 'Look, she's dying. Okay, she may die during the transplant or soon after. But it's the only chance she has.' It was shock therapy, and at that moment we accepted. We didn't have any choice.

And how did you feel, being given a death sentence like that? You say you're not a reflective person, but how did you take the news?

Actually, I was too ill to react. I was too busy trying to get better and I couldn't think very well. They said, 'You have to wait for a suitable donor organ,' and then nobody contacted me again. The waiting was very, very hard. I was also wasting away, because I couldn't eat. It was horrendous. Then they told me, 'If you fit oxygen in your house . . .' (which was a job because it was a house on several floors) '. . . you can go home on a Saturday and come back Monday. You can go for a little rest.' So I went. It was a big palaver with the oxygen machines, and they had to get me up the stairs with a wheelchair and all the oxygen . . . That was Saturday. Then on Monday, two o'clock in the morning, the telephone rang. Everybody was so fast asleep that no one else heard it, and I had to drag myself over to the phone. I picked it up and someone said, 'You have to come immediately to Harefield – we have some organs.'

I had left *all* my things at the Hammersmith. Everything was there, because I'd been in the hospital for months – all the cards and the flowers and the books, my nightie, shoes, make-up – everything was at the Hammersmith, but I never went back. I went to Harefield.

I'd worked with Sir Magdi at Harefield every week, so I knew the way. But Danny was driving, and he didn't have a clue. It was the middle of the night; my son Sebastian was in the car too, and they were both terrified. I was saying, 'Turn right, turn left, go straight.' We must have arrived at about three or four in the morning. They asked me for authorisation to donate my heart, and apparently I signed very happily. But I don't remember any of this.

Why did they give you a new heart as well as lungs, if your heart was good enough to give to someone else?

It's a simpler operation to take the whole lot out and put in another lot. The other way means more dissection.

But back to our arrival. Finally, a stretcher comes to collect me, and that was horrendous. I was frightened and sick; I was saying, 'Goodbye, goodbye' to the family, and I really thought it was goodbye forever. Danny saw another stretcher being carried in at the same time, and he saw that I had consented to give my heart. Then when I was in intensive care after my operation, he said, 'Don't tell anyone, but I think that might be the man who has your heart.' Later, when we were walking in the ward after we'd both recovered, the man said, 'You saved my life; I got your heart.' But he passed away about two or three months later.

After your operation, did they offer to show you your own organs?

No, not at that point. Because we were working together – the Hammersmith and Harefield teams – we had a bleep system to let us know when there were organs to collect from Harefield for our research labs. One day one of my colleagues was asked to collect a specimen of pulmonary hypertension. They all knew that I was waiting for a transplant, so he asked, 'Is that Julia Polak?' and they said yes. So he went to collect the organs and then dissect them. Did my colleagues find it difficult working with my organs? Some did, but apparently some not.

But afterwards I did see them, yes. Very interesting: they were lovely in terms of pathology. I presented them at the Hammersmith Hospital. They have a famous thing there called the 'staff rounds', where they present three cases: a clinician presents the clinical picture; if it's a surgery case the surgeon will talk about how he did it; and then a pathologist presents what they found, either in post-mortem if the patient died, or from the surgery. It's a regular event. This was an anniversary of some sort; I was at home recovering and they said, 'Can you suggest someone?' I said, 'You can present my case, if you like.'

It was a large lecture theatre, for about 1,000 people, but many people couldn't get in and had to be outside. Some people were crying – it was emotional, you know. [laughs] They sent a note round saying that I was very susceptible to infection, because it was still very early days. I didn't know about that note, but I thought, 'Funny, why don't people come and greet me?' Now I understand: people were told not to come near. They were all going . . . [she mimics them all waving from their seats]

My brother came from Argentina with his wife; Daniel was there; friends were there. Everybody was crying. For most people it was the first time they'd seen me since the transplant. It doesn't happen often, a colleague with a transplant! And, of course, I'm not very big, so I was just standing there looking small. The clinician who'd diagnosed my disease, Professor Oakley, presented; then Sir Magdi presented the operation; then I presented the pathology. My own lungs . . . Everyone was [she mimics everyone gasping]! The case was published in the BMJ [British Medical Journal] as 'The pathologist as patient'.

Did you yourself feel emotional at the time?

Well, no; I was trying to remain strong, because I was still quite wobbly, and I was busy trying to concentrate on what I had to present. And I felt strange that people were waving at me and they were all crying. [*we laugh*]

It is an amazing story.

At these staff rounds the pathologist has to conclude. Apparently I said, 'This pair of disgusting lungs . . .' It was an extreme case of pulmonary hypertension, probably the worst I have seen, and I couldn't imagine how I had managed to breathe through those lungs.

What relationship did you have with the material when you were presenting it as a case?

It was just slides. A fascinating case, but just slides. [*we both laugh*]

But people offered you all kinds of counselling support after the operation?

Well, they always do after a transplant, but I didn't take it. I did have nightmares for a while: I'd get up in the night saying I couldn't breathe. Of course, I could breathe, and in the end obviously I worked it out for myself.

You now have the reputation for being one of the longest-surviving recipients of a heart-lung transplant. What are the big risks?

You can die during the operation, because at certain points you are not alive: they take out your heart and lungs and put you in an artificial heart and lung machine. You can die when they put the organs back if they don't 'take': you need to get the organs pumping, the new heart that has just been connected up to your own blood vessels. And then you can die – which I nearly did – when they put you in this artificial heart and lung machine because all your blood clotting is altered. I was bleeding like mad, and Sir Magdi said, 'She needs fresh blood.' 'Fresh blood' means not from the blood bank, so they had to bring in the soldiers from the barracks near to Harefield. They didn't have time to test for HIV or anything else. I was *dying*. They tested for matching, and that's all. But my lungs, the tissue, didn't match. Sir Magdi explained the situation to Danny, but at a certain point you have to decide: do we match, or do we take the chance? You might find the perfect match if you wait, but it might be too late.

What do you have to do now to stay healthy?

I have to take drugs to suppress the immune system, to prevent rejection of my new organs. They aren't difficult to take, except that I have to take loads in the morning and loads in the evening. But the consequences of taking these drugs are not nice. You can develop malignant tumours. You can develop very serious infections because your immune system is suppressed. And some people die from what is called chronic rejection. Tomorrow is my 'annual' check-up actually, my MOT.

And do you get frissons of fear before your check-up?

Oh yeah, I'm completely neurotic. Every time I do the lung function test I hate it. My lung function is declining, and it's difficult to tell if it's age or just the lungs. But this is uncharted territory and you just hope . . . Nobody knows.

But how do you feel in yourself?

I feel fine. I do gym every day; I do yoga once a week, which I hate; it's so painful! I work all the time; I never stop. But you don't know what's next, and nobody can tell you, because they don't know either.

So do you feel vulnerable because your immune system is suppressed?

Oh yes, very vulnerable. It's not a good idea to go in aeroplanes because of the germs circulating in the air. I do fly, but we go first class, which has a bit more distance between people. Washing hands is very important, too. And I make sure that when people come to see me they are not unwell.

So tell me, what impact has your experience with this disease had on your career and on your life?

It has completely changed both.

When I got back to work after the operation, I thought I had to do something related to what happened to me, and I wanted to do more study on pulmonary hypertension. But then this new field started: tissue engineering and regenerative medicine, or stem cell research, creating lungs to overcome the lack of donor organs, etc. You could say that I would have gone into that field anyway, because I always liked to be into the exciting new areas. But it looks more glamorous to say it was as a result of my transplant. It's like the plot of a good novel . . .

So you left your original field?

I changed fields completely. Well, the purpose is the same – it's all geared to saving human lives. And the technology is the same, but the projects are different.

So you were a pioneer in the tissue engineering field. How would you describe your work to someone who isn't a medic?

Well, we all have the capacity to regenerate, otherwise you would cut yourself and you'd die by bleeding, but you do heal. If a baby has a cut it will not easily leave a scar, because a baby has great capacity to regenerate identical skin. But the older we get, the less able we are to repair damage well, and the more we get scars and marks. If we could regenerate really well we wouldn't get ill. If you cut off the leg of a lamprey or a frog, it regenerates very well. So the question is: what have we lost? Or what have we gained as an *inhibitor* of that process? And what can we learn that will enable us to get humans to repair like that?

Smoker's lung, for example, is a very common disease. The airway epithelium [lining] is destroyed because of the nicotine. Then the air sacs become enlarged, which is emphysema. Maybe we can encourage the lungs to regenerate, or we can ameliorate the condition, or stop it getting worse. There are several ways in which you could help the body to regenerate itself. You could grow a new lung *in vitro* [in the laboratory]. That's not going to happen tomorrow, but there are steps in the right direction. There are clinical trials on bladders that have been produced *ex vivo* [outside the body] – a whole bladder. Such an organ has been transplanted into children and they're still functioning eight years later. It's not a perfect bladder – they didn't do the ureters or the urethras – but still, it improved the condition of the children.

And these started with the child's own stem cells? So it's a match to the individual?

Correct. But one day maybe the famous (or infamous) embryonic stem cells, which are very 'plastic' and can become anything, could be used – if we can overcome the immunological problems caused by the fact that they don't come from the patient, and the ethical and moral concerns, etc. Umbilical cord stem cells are also becoming quite famous

– Richard Branson is starting an umbilical cord blood bank. Why? Because so far it has been proven in a *very limited* way that umbilical cord blood contains stem cells, like the bone marrow, which are efficacious for leukaemia, but for nothing else, so far. The cells are not like pluripotent embryonic stem cells; they are more like the bone marrow – they are multipotent. So they are more restricted in their ability to generate different cell types, but they have great potential.

Very recently it has become possible to produce pluripotent stem cells without destroying embryos. Adult cells can be taken from skin or other organs, and transfected with genes that will induce pluripotency identical to what's seen in embryonic stem cells. These are called 'induced pluripotent stem' [IPS] cells, and they were discovered by Japanese and American researchers. With more research, it may turn out that IPS cells can be used and that could obviate the ethical issues associated with the use of embryonic stem cells.

So there are several kinds of stem cells that could be useful. And the idea is: either you can have an *ex vivo* organ; or you could grow cells and then give them to a patient – that is, start the process of growing a new organ and then hope it will continue by itself inside the patient. Or you can kick the patient in the backside and say, 'Mobilise your stem cells,' and those stem cells migrate to the injured site.

And how would you do that?

You need to create drugs that will mobilise your own stem cells. It's early days, but it's done already in haematology, and probably will be done more widely in the future. So these are the three big approaches, and people are working all over the world on different strategies.

But there are enormous challenges. First, there is the blocking of embryonic stem cell research. Then there are problems with taking cells from a patient – they are older cells and they may all carry the same genetic defects, so they are not marvellous. Umbilical cord cells, as I said, are ideal, but there are not enough of them. So if you want to cure smoker's lung, what do you do? You need mass production, *industrial-scale* production, of things for clinical application. Nobody knows how to do it.

So all those challenges need to be overcome. I don't need to tell you all this; I could say it's all marvellous. Yes, that's so in a glass-half-full kind of way; we are certainly taking steps in the right direction – but my God, there are hurdles!

But at the moment have you got proof of principle that you can regenerate a lung?

No. Take my transplant, for instance. This lung belonged to another person, okay? I don't want to test it while I'm alive, but suppose I die, then my lungs can be stained and they can be checked to see whether they're being repopulated by my own cells. We did demonstrate and publish evidence that donor lungs can be repopulated by the patient's own cells. So there is an attempt by the body to regenerate even a foreign lung when the lung is damaged. Obviously, there is the release of something in these damaged lungs that attracts the cells to regenerate.

We are now working with experimental animals at Imperial College on a model of emphysema. We will put in umbilical cord stem cells, which are expanding, and see what happens. And if these cells do not produce any unwanted effects, and lung function improves, then we can consider going to a larger animal model and then man. So you see, the timelines of the conventional pathologist are totally different from those of research. A pathologist wants to save life now; they are aware of all the things that can be done, and they can suggest a course of action and see the result. But I might not see any of my research come to fruition in my lifetime.

Also, we must not lose sight of the fact that many important discoveries are made in this country but are commercialised everywhere else. Like monoclonal antibodies: this country didn't make a penny from their discovery, which is typical! That could happen to us in this field too, so we have had to make sure that we protect our discoveries.

So you've started a company?

Yes. I have two different things – one is my charity and one is a company. The company is going reasonably well, but it's early days.

The company is fully integrated with all the discoveries at Imperial, and I am the chairman of the scientific advisory board. We have a management group who are financiers, and they understand business. And then I run the science. And that's fascinating, because the science moves from the test tube to something that could become reality. Doing research and writing papers about your discoveries is beautiful, and you become famous. But it's real life that's important. In real life, you want to save lives . . . And to protect your discoveries, and make money to bring it back to the research.

Tell me about your trust, your charity.

When I was in hospital, and quite sick after my transplant, I was trying to collect money for the Harefield, and Sir Magdi said, 'No, set up your own charity, for your own research.' We started collecting and we supported a lot of research students, and now Imperial College wants to establish a prize for young women, called the Dame Julia Prize. We're collecting that money to help give the young women a chance.

Have you experienced barriers as a woman doing science?

I am very thick skinned, so I didn't. But lots of people say there are barriers. I do strongly believe we need to nurture women, because they lack confidence. We need to discriminate in favour of them. So giving this prize will help.

You say you are too thick skinned to have noticed any kind of sexual discrimination. Are you an ambitious person?

Yes, I'm driven. I am all the time organising and getting into new things. I mean, I will never be fully retired because I will always find other things I want to do.

You obviously have an extraordinarily busy life and yet you've brought up three kids . . . How did you manage it?

I have a good husband. And we had an amazing nanny. Danny and I had a rule that we never went away at the same time; one of us would always be here. And when the children were young, we were back home at 6 p.m. So it was a very structured life.

Golda Meir, when she was the Israeli prime minister, once told an interviewer that as a woman with a high-flying career as well as children you just have to come to terms with the fact that you will always feel guilty about something. Do you agree?

Oh yes, completely horrible! My son reminded us . . . Now he denies this story, but he was such a difficult person when he was a teenager. [*said with an indulgent smile*] He said, 'I'm writing a play to say goodbye to school; will you come?' We were so proud: the great genius writing a play, acting in it and staging it . . . Somebody said, 'I don't think you should come.' And we said, 'What d'you mean? Of course we'll come – all the family! Nanny and all her family too.' We were in the front row, and the story was about a disturbed child, who is disturbed because his parents are academics and flying all over the place, and he commits suicide in the middle of the stage because his parents forgot his birthday. We *never* forgot his birthday! We said, 'We're going to get out every single photo album with pictures of him blowing out candles at his birthday.' And then we couldn't find the damn albums. [*we both laugh*]

It's hard dealing with teenagers. I could have killed them all. [*laughs*] They're lovely now, absolutely wonderful.

And what about your Argentinian family? You say your grandparents came from Eastern Europe – under what circumstances?

You know, a lot of people emigrated at the turn of the century from Russia and Poland and places. I think it was just chance where they ended up. We grew up thinking of ourselves as completely Argentinian. But now we have been here much longer than we were in Argentina and the children are more English than Argentinian.

My family were lovely middle-class types. A very friendly, Jewish family. My mother was very intellectual; my father was a lawyer and then a judge. They were never practising Jews, so I don't know when our Jewish faith stopped meaning anything. We know it as our roots, that's all. My husband comes from a very similar background, but not wealthy at all. His father had a great intellect.

And how did your families fare during the dictatorships?

It was very hard. Daniel's brother, my brother-in-law, is one of the 'disappeared'. It was the mid-1970s. He was taken away in front of his father. They said, 'If he has done nothing he will come back.' He had done nothing, but he never came back. They most probably threw him in the Atlantic, and they never recovered the body. I think when he was taken away it virtually killed his father – he died some years later, but I think he died of a broken heart.

Culturally, as Latin Americans, how did you find working with people here?

Everything is so civilised here, and the culture . . . I mean we're spoilt here; it's amazing.

Finally, Dame Julia, what specifically are you working on at the moment?

Right, the biggest challenge at the moment . . . There is lots of research being done all over the place on cultured cells two-dimensionally in a Petri dish. But we are three-dimensional beings, so therefore we need to learn how we can culture in three dimensions – because the cells talk to each other, and talk to the surrounding tissue.

This is another area where we have strong collaboration with engineers, because we need to grow the cells in what is called a 'bioreactor'. The best example of a bioreactor is the womb. The development of the baby in the womb is controlled by all these natural signals in the three-dimensional environment. We need to grow the cells in this kind of way, and people can't do that at the moment, so that's what the challenge is. Not mine alone, but for everyone working in this field.

As far as you're concerned, is this the future?

I think regenerative medicine – in my unbiased opinion – will rewrite the books of medicine. Of course! I mean, that's where we should be going.

'CHILDREN ARE NOT JUST LITTLE ADULTS'

Miguel Reyes-Múgica
*Chief of Pathology and Head of Laboratories, Children's
Hospital, Pittsburgh; Professor of Pathology and Marjory
K. Harmer Chair in Pediatric Pathology, University of
Pittsburgh*

Miguel Reyes-Múgica grew up in relative privilege in a small town in Mexico, where both his parents were doctors. As a child he would accompany his father on visits to patients in rich and poor homes, and his experiences kindled in him a strong social conscience and a desire to follow in his parents' footsteps. At medical school in Mexico City he fell under the spell of Dr Ruy Pérez-Tamayo, one of Latin America's leading intellectuals. 'He was professor of pathology, and when I took that course in my second year I immediately knew I was going to be a pathologist. I wanted to be just like him in many respects,' he says.

Another big influence on his career was the earthquake of 1985 that levelled much of Mexico City and killed many of his colleagues. When the general hospital where he was about to start work was threatened with demolition, he went instead to the National Institute of Pediatrics, one of the largest and busiest paediatric hospitals in Latin America. It was a time when paediatric pathology was gaining recognition as a vital specialisation in its own right.

Reyes-Múgica went to the United States in 1990 for a year's research, but for personal reasons never returned to work in Mexico. He went to Yale in 1994, where, as director of the programme in paediatric and developmental pathology, his personal focus was childhood tumours and melanocytic lesions (giant pigmented birthmarks with a potential to turn malignant) in children, the latter sparked by an extraordinary case he had seen in a baby in Mexico.

———•———

I was very close to my parents, and particularly to my father. He was more of a friend than a parent at times. He was a well-read man and, although not very religious, we would speak about life and death, and he helped me to discover my own path in medicine. At that time there were no structured programmes of specialisation in Mexico, so he was a generalist, but a very knowledgeable generalist who would take on any kind of case. My mother was more surgically oriented, so they complemented each other. Her hospital was at San Cristóbal de las Casas in Chiapas, where the Zapatista Revolution started in 1990. I grew up mostly in that small town, although I was born in Mexico City.

What sort of social environment was it?

There are many indigenous people there, and the relationship between the more Spanish-looking people and the indigenous natives has never been easy. But my mother and father would see patients from all strata, from the well-off to the very poor people – frequently, you know, payments were in the form of a chicken, a bag of eggs, fruits or things like that. My father labelled me a communist when I was about 11! I guess I developed a little social conscience when I saw the differences in the standards of living.

I was expelled from my high school for rebelling when I was 15, and my father put me to work with his parents, who ran some butcheries in the local market. I used to go with my grandfather to buy cattle, and we would slaughter them, and then we would do a complete post-mortem. So my first contact with cutting and looking at flesh really was there.

You say 'post-mortem', but, I mean, butchery is a lot different, surely?

Oh, a post-mortem in a human, an 'autopsy', is absolutely different from butchery. But it was an animal: it was muscle, it was viscera, it was handling a knife, it was knowing the different types of tissues from a macroscopic, naked eye, perspective, and recognising when something is not how you expect it to be. I recognise that experience as a very important influence on my personal development in pathology, and I can tell you that when I cut an organ, not only in autopsy but also in surgical pathology – which, by the way, is what we do most of the time; we are not 'death doctors' – the residents usually express surprise at the way I handle the knife.

And then you tell them, 'Well, I learnt on cattle'!

Yes, exactly. Even though I am a paediatric pathologist and usually handle smaller organs, I use a longer knife, and I can cut very thin slices without injuring myself. I am a very neat pathologist.

So how did you not end up becoming a master butcher?!

Well, I had to go back to school. So my father and I made a deal and I went to a military school in Mexico City, where I spent a year fighting my way out of little troubles every day. After a year I left and went to a more normal high school, and then I entered medical school. But I learnt discipline in the military school, and it was a useful exercise. For a youngster from a relatively privileged family it was a good moment to realise that things can get really tough, and if you want to make it you really need self-discipline.

Because things happened very fast in my life. First I got married, aged 18, and the same day that my first daughter was born I was accepted at medical school – in the largest university on earth, which is the National Autonomous University of Mexico. The first day of class was interesting: I was the only one with family responsibilities. I was very fortunate because I had four or five wonderful classmates, all of them very intelligent, and I became part of this group that would get together to study. I had much less time than they did, because I had to earn a living as well. The first year of medical school I was assistant to a producer of a TV programme on cultural matters, so I had to read a lot in addition to reading my medical school books. It was fun, it was a lot of work, but when you are young you can afford not to sleep much!

But television started to take second place pretty fast. After the first year I realised I had to concentrate on medicine, so I looked for a position in the medical school as an assistant. They opened a competition for people that had finished some first-year courses to start teaching those courses to the following class. So I took the competition and I won two places. The competition was very tough because my class was tremendously big, about 5,000 students. They selected the best 100 students and, out of those, six won positions to teach – and that's when I discovered histology.

The first histologist was a Frenchman, François Bichat, in the 1700s. He never used a microscope, but he described 20 different kinds of tissue just with his naked eye. He was the 'father of histology'. When I started looking down the microscope, I discovered a different universe. Those years were very important in my life, because I had such a need to make money, to concentrate on my career, and I discovered that teaching something that was part of my own academic learning was the best combination possible. In the mornings I would teach histology, and in the afternoons learn my own subjects in medical school.

Then I discovered several mentors, but one of them, Dr Ruy Pérez-Tamayo, is the most important person in my academic life. He was professor of pathology, and when I took that course in my second year I immediately knew I was going to be a pathologist. I wanted to be just like him in many respects. He's now in his eighties. He has shaped the discipline of pathology in Latin America, but he has also made tremendous contributions to pathology worldwide.

You say that when you first looked down a microscope it was a different universe – what was the thrill?

Well, the first time our lecturer told us, 'Look at these cells on the screen' – they used to project these slides – I couldn't really understand it: what were cells, what were nuclei? So it was a challenge, and I like to be challenged: there were a whole lot of things I wanted to know. But then I discovered I had some ability to distinguish things under the microscope, and I liked that. It gives you immediate gratification to be able to diagnose something under the microscope, be it normal as in histology, or abnormal as in pathology. Pathology is just ordinary life in abnormal conditions, right? So when you know your 'normal', you are able to recognise when something is not normal, therefore pathological. And microscopy is magnificently beautiful! I still now can spend more time than I should just looking at something under the microscope because of the beauty of it.

So Dr Pérez-Tamayo was your mentor at that time? And did he take you under his wing?

Well, I took his course and he was aware of my presence relatively fast because I asked questions and things. Every year he would select one student to go and visit his lab, and that year he invited me. This was a completely different universe again. Here it was not a matter of looking down the microscope; it was doing all kinds of strange things that scientists do in labs. He was an expert on collagen, which is the most abundant protein in our bodies. I started working with his group on extracting collagen, and trying to identify an enzyme that would degrade the protein in order to explore the biology of certain diseases such as liver cirrhosis.

I spent about 18 months learning electrophoresis, protein extraction and things like that. And in the process I learnt many other little lessons in science – experimental design, and to be sceptical, never to believe the first time you see something.

So did you decide then that the academic side of things was more beguiling to you than actually seeing patients?

Well, yes, but I never gave up my intention of seeing patients, because when I started with my father I discovered I had some ability to relate to patients and their families, and I liked that experience. Even now I sometimes go to the wards and look at patients myself when I have specific questions to answer. Pathology is not only research; it's not only looking at microscopes and glass slides. I never forget that behind everything we do there is a patient. So I am a doctor: I consider myself a physician with a particular subspeciality.

After working with Dr Pérez-Tamayo, how did your career develop?

I continued my medical school studies. I finished the four years of basic learning and then I did one year of practice, the internship, and then one year of social service. This is when you pay back society that has been generous enough to provide you with a free medical education. I worked at the National Institute of Pediatrics in Mexico City, and because it was one of the largest and busiest paediatric hospitals in the whole of Latin America, I rapidly began to develop some knowledge of pathology, in particular paediatric pathology. After my social service year I started my residence in pathology, and the director of the programme was again Dr Pérez-Tamayo, so I went back to him. I eventually went to the General Hospital of Mexico, the largest in my country, where I finished my specialist training.

Then there was another significant event in my life, in *all* our lives: the big earthquake in Mexico City in 1985. I lost seven of the residents in my department, including my room-mate, in that earthquake. Together with other people, I pulled him from the rubble of the collapsed building. Forty-nine residents just in my hospital were dead . . . Many people died.

Where were you when it happened?

It happened at 7.19 a.m. on 19 September 1985. I was kissing my small son, and we both fell to the ground with the first shake. Mexico is an area with high earthquake activity, and I have many memories of my father sitting on the bed looking at his watch and counting the seconds, and my mother praying. I'd had that experience many times, and I was never very scared. But I was a little surprised that this one was so strong.

I was living with my three children at the time – I had separated from my wife, and I kept my children; I raised them. So I asked the nanny to take my son to kindergarten, and I took my two daughters to school. I could see people running along the streets, but I didn't see collapsed buildings – until I got to my hospital. Then I began to realise the magnitude of the problem. Two buildings within my hospital collapsed – one was the building where residents lived, the other was the gynaecology/obstetrics building where the nursery was. There were 302 people killed in my hospital alone. The official death toll varied tremendously, but probably about 20,000 people died in that earthquake.

The army took over the hospital and the city in general, and we organised groups to identify people pulled from the rubble, and tried to help the injured. We would spend hours trying to remove rocks and material. There would be 50 or 60 people working in an area, and someone would suddenly yell, 'Hey, silence!' and you would hear someone asking for help. There were people, including newborn babies, pulled out several days later still alive. It was a terrifying experience.

What effect did the earthquake have on you?

It changed my life in many ways. First of all I am terrified of earthquakes now; I learnt my lesson. Secondly I switched my decisions. I was going to be chief resident in the General Hospital of Mexico that year. But after the earthquake there was talk about closing the hospital because of the damage. I switched and took a job in the National Institute of Pediatrics [NIP], where I'd done my social service. And instead of being a chief resident for a year, I began, in 1986, as a fully fledged junior staff pathologist in a very big hospital.

What sort of cases were you seeing?

NIP was a tertiary level centre with the ability to perform renal transplants and cardiac surgery and stuff. But also we had a very large caseload of run-of-the-mill infections, and the pathology of poverty – malnutrition, tremendously advanced cancers, all kinds of things. It's a hospital that covers the full spectrum of human pathology, from zero to 18 years of age, so I was exposed to a massive amount of paediatric human pathology in the four years that I spent there.

How did you manage the two things – your career and bringing up a family?

Well, my ex-wife had some psychological problems. We decided to split and I kept my three children. I was in the second year of my residency training when the divorce happened, so it was tough. Usually I would take the children to school in the morning, and I would take an hour off to eat with them around 2 p.m. At 3.15 p.m. I would run back to the hospital and continue my activities, and then in the evening I would go back to the university to teach. So it was crazy: I didn't have much time to be with my children, so I had to spend what is now called 'quality time' with them. And eating is an opportunity to exchange a lot of things; it's a very important time of the day, and an educational experience for children.

And did you remain close to them?

Oh yes; they lived with me until they all left home. They are married now, and I have one granddaughter.

When I divorced I met my second wife, who was a medical student. It was tough, because she came from a very conservative family that wanted her to marry someone without a lot of baggage, and I had three children, but she finally married me. After four years together in Mexico we decided to try to come for one year to the United States.

I went to speak with Dr Pérez-Tamayo and he mentioned González-Crussí, who was a huge figure for me. He was a fantastic paediatric pathologist, a famous Mexican, and he had written *Notes of an Anatomist*. So it couldn't get better. Dr Pérez-Tamayo wrote a letter and a week later I got a phone call from Dr González-Crussí inviting me to visit Chicago. He said, 'If Dr Pérez-Tamayo recommends you, you're accepted'!

Your intention was to come and do some learning here and then go back home?

Absolutely. I never even took the United States Medical Licensing Examinations [USMLE], so I wasn't officially allowed to take part in any medical procedure because I had no credentials. I was a research associate doing projects, and I was just learning pathology on the side with Dr González-Crussí and his team. But then things got complicated with my ex-wife and it was very difficult. I sent my children on vacation to Mexico and they were kept there against their will. Finally I recovered my children in a complicated transaction that took all my little savings. My wife was pregnant at that time, and we decided to stay in the United States and cut our ties with Mexico at that moment.

Dr González-Crussí told me, 'If you can take the USMLE, get your credentials in time to be appointed fellow, I will keep the position for you.' I got my licence just a week before the deadline and I stayed on in Chicago as his fellow in paediatric pathology.

This was already my fourth year in Chicago, and then I saw that Yale University was looking for a paediatric pathologist. That was in 1994, and I was there for fourteen years.

Tell me about your time with González-Crussí. You had started reading his essays before you met him, had you?

I have read every book he has published to date. Dr González-Crussí is a vastly cultured man. He speaks many languages. But my first impression was that he was a fantastic pathologist. Pathology is a difficult trade; you need to be special in certain ways. It's like, to do basketball you need to be tall. Well, to be a good pathologist there has to be something in your brain that allows you to orient yourself in the visual, spatial field, and he had a particular talent with that. But also massive medical information – he knew everything, and I was very impressed with that. Then I started reading his contributions to medical literature, and saw that he was able to jump from one topic to another in paediatric pathology, always as if he was an expert in that particular topic.

In his writings Dr González-Crussí has considered death a great deal. What are your own thoughts about death and dying?

I'm not a religious person, so I have tried to look at the process of death from the most scientific angle. But even so I recognise there are certain things completely beyond our scientific understanding. And I have had my own personal experiences – I don't make too much of them, but others perhaps could.

I was called to do an autopsy on a baby who had a disease that required very rapid intervention in order to make the diagnosis, and to provide genetic counselling to the family. The only option was to go immediately after her death into her heart and take some parts for examination. The clinician called me before the baby died, asking, 'Would you be available if she dies in the night?' He called me later and said, 'We're going to interrupt artificial life support in half an hour, okay?' So I drove over in the night, and they came with the baby a few minutes after she died.

When you die, cells are still viable in many organs, and when you stimulate certain tissues that are contractile they can develop mechanical activity. And what happened was that I was doing the autopsy with the head of the paediatric intensive care unit at my side, and when we got to the heart it started moving. The baby was dead, but the heart was sort of beating . . . Not really, but in a disorganised fashion there was a beat. I felt at that moment like an Aztec prince taking out the heart of a human sacrifice, if you will. And I wondered at that moment, 'What am I doing here, opening the heart of this little baby?' Of course, that was just for a flash second, and I understood that what I was doing was very important. The diagnosis was made, and the family now knows what type of disease they might carry, and understands what to do. It was a very valuable autopsy. But the experience was very chilling.

Is it something that has lived with you?

It has. It has. You wonder . . . In many autopsy rooms there is a statement in Latin that says something like, 'This is a morgue, the place where death helps those that are still alive.' And that's exactly what came to my mind – the contribution this baby was making to her family and to humankind was huge. So I see autopsy as the most exhaustive, and final, medical exam that a doctor can provide for a patient. I don't see autopsy as something horrendous and macabre.

But it was just too close to the event of death – was that what got through your professional defences?

It was too *different*. When you do an autopsy, usually the body has been in the refrigerator for several hours, it's cold, it has changed colour and there are many familiar things. But with this little kid I saw something different. She was dressed as a baby. She was pink and warm. It is very difficult to open a body like that – and just very different.

People can lose track of what life is all about. Frequently I see a case where the clinicians have spent weeks, or sometimes months, treating a patient. They know all the parameters – respiratory, cardiac, etc. – and they follow those and watch the curve of progress up, or down. Then finally, when the patient dies, they come to me and say, 'What happened? We don't understand how this patient died.' And my answer on several occasions has been, 'Well, I can't understand how this patient *lived*.'

Our practice of medicine nowadays is so fragmented into different specialities and little tasks that we as doctors lose track of the patient. So they come and say to me, 'But this patient was doing fine. The blood pressure was this and that, the urine was so . . .' But when I just show them the heart, the lungs, the whole thing . . . I mean, you can no longer recognise the human body. The baby weighs two or three times the normal weight because it has been inundated with fluids. It has seven or eight catheters in different places. It has been operated on three times . . . What I'm trying to say is that medicine is now a practice in which lots of specialists lose track of the big picture, and frequently it is only the pathologist that can try to bring all the loose ends together.

There are situations in which specialists put too much trust in their results. They are not sceptical enough, and they make mistakes. A memorable occasion was when a genetics counsellor stormed into my office very upset because I had issued a report saying, 'Fetus without significant abnormalities' after an autopsy was performed. She said, 'What do you mean? We saw polydactyly in the ultrasound; that's why we interrupted the pregnancy.' Polydactyly is more fingers or toes than five. To interrupt a pregnancy on the basis of polydactyly may or may not be justified, but that's not the issue. The issue is that I knew the specialist was going to challenge my findings, so when I performed the autopsy I put both hands on a glass slide – they were so small that they could fit into this little space – and you could see five fingers on each hand. So I didn't open my mouth; I just took the slide and gave it to her in response to her question.

The point is that with ultrasound – as much as it is very advanced and you can trust a lot of the results – you should never lose track of the fact that those are virtual images, and pathology has *real* images. This is where things stop: there is no more accurate and real exam than a pathological exam.

Do you feel people are relying too much on technology, almost in order to cheat death?

I recognise that if we do not apply our marvellous technology to these very sick patients, we would not be making the progress that we are making. There are voices that argue very strongly against the application of advanced technology to sustain life, to treat certain conditions in childhood or adulthood. I disagree with that because, as you know, our quality of life is much better now than it was just 20 years ago. And 50, 80 or 100 years ago, when there were no antibiotics and no anaesthesia and things like that, if the doctors of the time had not applied all their efforts to pursuing a particular question, we would never have got to this point. So I think that although the end does not justify the means, we should have the curiosity and the will to apply this marvellous knowledge. But we should do it in a respectful way.

Tell me a bit about your own research.

Well, I have been interested in a particular group of disorders that involves the development of the neural crest. The neural crests are transitory structures of the embryo that are the excess tips of the closing borders of the neural tube. The neural tube gives rise to the brain and to the spinal cord. But as the neural tube closes – because it first begins as a groove – the excess tissue remains there instead of disappearing, and it migrates to different parts of our bodies through a very complex process.

Neural crest disorders are of different types. There are neural crest diseases that are cancer-like, or cancer. And there are neural crest disorders that are the result of an abnormal migration or survival of the cells. The bowel, for example, has a brain of its own. The peristaltic movement of the bowel is coordinated essentially by this brain in the gut, and this nervous tissue, which has many millions of neurons inside, requires for its development the migration of the neural crest into the bowel, and the proper development and survival of the cells in a coordinated fashion. If you look at it under the microscope, it looks like a little piece of brain sandwiched between layers of other tissue in the bowel.

That's amazing! And is there a lot of it?

Yes. It continues from the oesophagus all the way down to the exit. And the proper development of this gut brain is necessary for our bowels to move. There is a childhood condition called Hirschsprung's disease, in which babies, usually, are unable to move their bowel, and become distended. And they can explode, literally, with a colon like an anaconda. Just imagine what it is like not to be able to move your bowel for five, six days. There is no peristalsis, because there is an area that is lacking these neurons in the bowel.

The disease was described in 1888. It was poorly understood for the first 50 years and then it began to be unravelled by a surgeon at Boston Children's Hospital. For the first 50 years the surgery performed on these patients was wrong – they removed a normal piece of bowel and left the abnormal piece because it looks deceptively normal. But then we began to understand that, and the biology of this disease has been unpicked over the last 15 years. I have been more active in determining how to diagnose this disorder than in trying to understand the genetic basis, because for paediatric pathologists this is a serious daily problem. Frequently a biopsy comes to your table that says, 'Please rule out Hirschsprung's disease,' and it is very difficult to make the diagnosis, because it is based on the *absence* of neurons, and absence of proof doesn't mean proof of absence.

And is this a common condition?

It occurs in about one in 5,000. But the thing that I am more interested in is a little more uncommon. It's called neurocutaneous melanocytosis. This is a disorder of neural crest cells that produce melanin – so-called melanocytes. All the pigment in our skin, melanin, is produced by these cells, which all come from the neural crest – with the exception of those that are in the eye. Every other melanocyte in our bodies – including those in the brain – comes from the neural crest. There are babies born with abnormalities in the population of melanocytes, and they develop this condition, neurocutaneous melanocytosis, where they have extensive areas of their bodies covered by moles, or 'naevi'. These darkly pigmented areas can cover portions so extensive that they are sometimes called 'bathing trunk naevi'. Or they look like Dalmatians – hundreds, thousands, of little satellite lesions here and there. And the problem with those situations is that not only is the skin abnormal, but the deeper tissues are also abnormal and can develop malignant tumours, melanomas, inside those naevi. Frequently they develop a naevus inside the brain, and that can interfere with the development of other structures. They have seizures, they have hydrocephalus – dilatation of the ventricles of the brain – and frequently they die. The condition is relatively rare, though I can't give you an exact figure. One in 20,000 newborns will have a giant naevus, and some of those will be neurocutaneous melanocytosis.

But just over 10 years ago, I came in contact with a newly developing support group of parents. They had each had a child born with a giant naevus, and the doctors had freaked out and said, 'I've never seen this. I don't know what it is.' Or, 'It's a melanoma; let's treat it like this . . .' There was very little knowledge. So parents created this family support group. All of the members are either parents of children with these lesions, or they are patients themselves, because many people have survived 20, 50, 60 years suffering with this situation, and they are 'freaks'. Sometimes people can hide it. But other times, half your face may be black with a tremendously disfiguring lesion.

So that is where I concentrate most of my research efforts. I was doing that with my late wife – she passed away five years ago.

What are you discovering?

First of all that this is much more frequent than we used to think. People are now aware of this group and are coming forward and saying, 'I have one of those. I was born with this naevus.' We are also discovering that it's a very complex disorder in terms of genetics. We know of no twins, for example, that have this problem, so there is not an inherited basis. It does not repeat in families. There is nothing the mother did during the pregnancy that would represent a particular risk.

So we are first eliminating a number of things. We are sure that these are real clonal lesions – by which I mean lesions that arise from a single cell that starts to reproduce and forms a clone. But we need to gather a lot more samples and make a very organised effort in our research. And when you don't have such a massive number of patients, it's very difficult to gather data and get researchers interested.

What made you particularly interested in this condition?

Well, when I was a pathologist in Mexico there was a baby that came to the department of oncology, and it immediately became the 'patient of the month' – everybody was talking about it. The baby was born with something in the genital area that was very striking – they couldn't even tell if the baby was a boy or a girl. They thought it might be a tumour that would kill the baby rapidly, so they took a biopsy. What I saw was something *very* strange: it reminded me of what is described in congenital naevi, and at the same time it had features of nervous tissue.

Nervous tissue? From where?

From the neural crest. Melanocytes, in my opinion, are really modified neurons. They are cells that become specialists in synthesising pigment, but they have these dendrites, these prolongations – they are like an octopus, literally. They have cytoplasm that is the body of the octopus and they have all these 'tentacles' that carry the melanin and transfer it to surrounding cells.

I never stopped thinking about that case. When I moved to Chicago I told Dr González-Crussí about it and he said, in his classic reflective way, 'Very interesting . . . I have never seen a case like that.' Then a couple of months later he called me up for a frozen section analysis. They were operating

on a patient with a melon-sized tumour in the perineal area
that was surrounded by a patch of hyper-pigmentation, like
a naevus, and I remember his voice on the phone: 'Miguel,
you remember the case you told me about from Mexico a
few years ago? I think we have something very similar.' Sure
enough, it was identical under the microscope. So we decided
to publish those two patient cases together, in the oldest
pathology journal, Virchow's Archive. From that point on I
was hooked into the biology of melanocytic development.

*Dr González-Crussí was telling me that paediatric pathology,
as a specialism, is relatively new. Were you aware of that
when you came into this field?*

Yes, I was aware. I remember that when I went to Yale,
some of my colleagues were wondering, 'Why do we need
this guy? We can do the paediatric cases.' It was a fantastic
department at the time, and people were very well trained
and experienced in many fields – but paediatric pathology
is *completely* different from adult pathology. Under a
microscope, tissue from a child that is one day old is different
from the tissue of a child that is one month old, and that's
different from tissue of a one-year-old, and that's different
from the tissue of an 18-year-old. In paediatrics the key is
development. When you take a case to an adult pathologist,
the questions he or she will ask when looking at a slide, of
a tumour, for example, are essentially, 'Where is the lesion?
How fast did it grow? How big is it?' The first question a
paediatric pathologist will ask is, 'How old is the patient?'

So yes, when I came to paediatric pathology, I think
people had started to recognise the need. There had been
paediatric pathologists, pioneers, but as a speciality it became
more recognised around 25 years ago.

Tell me, how important is your profession to you on a personal level?

Well, pathology is a very important part of my life, but it's not the only one – and it's probably not the most important one. A friend of mine, another doctor who was with me in Chicago, said something that has remained with me. He said, 'You know, before I see you as a doctor, as a pathologist, I see you as a parent.' And I think that is what defines me: I am a father. Three of my four children live close by, my youngest lives with me, and we all get together every other week. We drink a bottle of wine; we cook and have fun. We speak about matters, we travel together.

That brings me to one final question: you didn't go back to Mexico?

I couldn't. The rupture with my ex-wife was so difficult that when my children were kept against the law for a few days in Mexico, I realised I had to keep a distance, for their health and for their education. So even though my original plan was to be a pathologist in Mexico after some speciality training here, I couldn't. I remained here and tried to make the best of it, but I have tried to give something back to Mexico, and to Latin America.

READING THE BONES

Sue Black
Professor of Anatomy and Forensic Anthropology,
University of Dundee

Sue Black grew up in idyllic surroundings on the west coast of Scotland, where her parents managed a hotel. An incident during a dustmen's strike when she was a child had a critical influence on her life and career. She watched her father beat a rat to death with a stick as it rummaged in the overflowing garbage bags behind the hotel. 'I could see its tail lashing, I could see its eyes, and I could hear it growling. And from that point onwards I've had an absolute and utter morbid fear of rodents,' she says. It even determined the choices she made in studying anatomy at university. Today Sue Black is one of a tiny community of forensic anthropologists in the UK. An expert at Disaster Victim Identification, she has worked for the International War Crimes Tribunal in Kosovo, the United Nations in Sierra Leone and the UK government in Iraq, and following the 2004 tsunami in Thailand.

Sue Black lives with her family in Stonehaven, Aberdeenshire, and admits it is sometimes hard to balance what is often a dangerous, though compulsively fascinating, job with the responsibilities of motherhood.

When I went to university I had no idea what I wanted to do eventually, except that it had to be something vaguely biological. At the end of second year the only two subjects I was any real good at were anatomy and botany. I went to see both tutors, and the botanist – bless his heart – was the most boring person on earth. I thought, 'I can't name and draw plants for the rest of my life. I *can't*! I'll do anatomy.' So I went into anatomy.

The third year was dissection and I absolutely *loved* dissection. But in fourth year you had to do a research project, and they all involved things like 'lead level in the rat brain', 'carcinomas in the hamster pituitary' . . . and *nothing* could persuade me to lift a dead rodent out of a bucket. It's a complete and utter, illogical fear because of my father. So I told the tutor, 'Look, I can't do mice, rats, hamsters – can't do them alive or dead.' And she said, 'Well, we can put a project together on human bone; how about that?' Perfect! As long as it didn't involve a rodent, I was happy.

So I did my honours project in the identification of human bone. Then my head of department, John Clegg, said, 'We've got some money if you want to do your PhD here.' So I've fallen into it the whole way along – which is a nightmare for any school that's trying to use you as a career model!

I did my PhD, and then a very dear friend of mine, Louise Scheuer, contacted me to say there was a vacancy in the department she worked in at St Thomas' Hospital in London. It was a fairly aggressive interview panel, but there were two people who wanted somebody in the post who could teach anatomy. At that time there were so many people in anatomy departments who couldn't teach anatomy: they were cell biologists, biochemists, etc. The head of department then was Michael Day and his final question to me was, 'If I needed you to go into my dissecting room this afternoon to teach, could you do it?' I said, 'Yes, of course I could,' and that sealed it.

So I started lecturing at St Thomas'. One day Iain West, the forensic pathologist, phoned the anatomy department and said, 'I've got some bones; does anybody up there know anything about bones?' I was sent down, and the most *miserable* policeman was standing there. He looked me up and down and you could see him thinking: 'Slip of a girl, what's she gonna know?' But we took the bones and put them in a plastic bag. They'd been found in a rubbish tip and were suspected to be a missing person. We put the plastic bag on the radiator and left the bones there for about 10 minutes, then I opened the bag and stuck it under his nose and said, 'What can you smell?' He said, 'That smells like roast lamb,' and I said, 'Exactly. They're sheep bones.' And he was so impressed, this policeman, that he'd got it right that next time there were some bones he said, 'Oh, I'll have that woman from anatomy.' So I just started doing more and more of the bones work around London, and then ended up doing work for the Foreign Office. It just sort of spiralled from there.

And what did you learn about bones? What can you tell from bones?

When you're given a pile of bones – it might be something to do with the World Trade Center; it might be the London bombings – the first thing is: are they human or not? It's easy to tell if you've got a skull. But if it's a tiny bit of bone from a finger or something . . .

You've got to bear in mind that with things like the World Trade Center, there were restaurants, so there was beef, pork, lamb in the remains. When the London bombs went off there were people carrying shopping – you know, they had Sainsbury's chickens and things. Or there would be cats or dogs in the tunnel. So you've got to separate out: is it human or not? Once we've established it is a person, we might be asked, 'How long has this person been dead?' Because if it's more than 70 years before the current date, then it's no longer a forensic case, it's technically archaeological.

Literally? That's the cut-off point?

Yes. It's man's 'three score years and ten', and if it's a murder case the chances of the perpetrator still being alive are, of course, slim, so there's little technically for police to investigate. You will always get cases that won't fall neatly into that category. For example, if you find children's remains on Saddleworth Moor, then it doesn't matter whether it's 100 years from now, they will still possibly be the Moors Murders. Certain cases have a notoriety.

So, is it human? How long has it been dead? And then, what more can I tell? Are you male or female? How old were you when you died? It's much easier to assign an age to a child than to an adult, because children go through a phase of quite regular growth, so that you can go into Marks & Spencer and buy a pair of trousers for a six-year-old. You can't buy them for a 42-year-old. Because growth and age are so closely related in a child, we can get very close. With a fetus, you can identify age to within weeks. With a young child it's to within months, and with an older child it's to within a year or so. By the time you reach puberty it's to within a couple of years. But once all the growth changes have stopped, then the human body goes through a stage of maintenance, in the twenties. So if a body is in a maintenance phase we can say that person is in their twenties. But beyond the twenties – unfortunately it seems very young – everything's degenerative. And some of us will degenerate quicker than others, so it becomes very unreliable to assign an age if you're over 30.

So, we have to assign a sex; we have to assign an age. We can then assign a height. Height's not very useful for separating people unless you're exceptionally tall or exceptionally small. Then the fourth indicator of biological identity is your race. But race is such a contentious issue, for a number of reasons. It's also a fact that we have such an admixture between the races that it's very, very difficult.

So that's the first thing we'll produce: a biological profile that says, he's male, aged between 25 and 30, 5ft 6 to 5ft 8 tall, white. Then you want to establish the personal identity. What information can you take from these remains that will separate two individuals with an identical biological profile? When it comes to DVI – Disaster Victim Identification – we have four principal means of identification: dental work, DNA, fingerprints and any unique medical condition, such as a hip replacement or a pacemaker with a unique serial number.

But that's not really forensic anthropology. Dental records are the odontologist; DNA is the forensic biologist; fingerprints are the fingerprint officer, and unique medical conditions are the pathologist. What does that leave for the anthropologist? In many ways we get the scrapings at the bottom of the barrel. We know our position! That is, if you can't get identity by any other means, come back to the anthro.

So do you generally work as a team with these other people, or are you just called in after everybody else?

Depends on the situation. If we're working on a deployment for DVI, we will be part of a team. If it's a case where the police bring in a bone to you, then you're on your own, because basically they've decided that pathologists can't do anything; they can't get any DNA out of it. And so it's all about trying to establish biological identity.

For example, we had a case in Scotland where a middle-aged woman went missing, and her husband's plea was that she'd gone down south to support a friend who had marital difficulties. But the trouble was that this woman, every night of her life, had phoned her elderly parents at the same time, and she'd stopped doing it at that point. That change in behaviour is an indication that something's wrong. So the 'scene of crimes' people went to the house; they found some blood in the bathroom; they found a chipped piece of her tooth in the U-bend of the bath. But that doesn't mean she's dead. She could have gone into the bathroom, tripped, cracked her chin on the bath . . . But they found her blood on the door of the washing machine, and in the filter they found a tiny fragment of bone no bigger than about 10mm long, maybe 4 or 5mm wide. And that's all they had. DNA showed it was this missing person. But the question was: which part of her is it? Because if it's a bit of her finger she could still be alive, but if it's something more critical, then we're in a different story.

We could identify that that tiny fragment came from the left greater wing of the sphenoid bone, which is around your temple. That's the only place in the whole body that fragment could come from. So then you can confront her husband and say, 'This is a bit of her skull, and it's found in the washing machine . . . We need an answer.' He changed his plea. He said that they'd had an argument, she'd run out the back door, tripped on the top step, cracked her head on the patio and died. He stated that he'd picked her up, which is how her blood and bone got on his clothing, put her in the bath, which is how her blood and her tooth got in the bath, wrapped her in plastic and dropped her body in the local river. We've never found the rest of her body. All we've ever had of this missing person is this tiny fragment of bone.

And was there a conviction?

Absolutely. The pathologist's testimony in court stated that it couldn't have been a single blow because the bone fragment was dislodged on to his clothing, and when he put his clothes into the washing machine, that's how the bone got into the filter. He was convicted of manslaughter. So we've no idea, when we get a tiny fragment of bone – it may go absolutely nowhere, but it may lead to a conviction for manslaughter.

When you started in this field, were there people with this kind of experience who could teach you, or have you pushed at the boundaries of knowledge as you've gone along?

A bit of both. There's always been a very good relationship between anatomy and forensic matters, but it was never a formal relationship. My PhD supervisor was interested in bone, so that was useful. And she was an exceptionally good anatomist, so we kind of learnt together and tried to keep up with the latest developments. Then when I moved to St Thomas' Louise Scheuer was there, who's also an *exceptionally* gifted anatomist who was interested in bone. So I always had strong women around who had the kind of information I could 'feed off' and develop. But there was no formal training; you couldn't do a degree in forensic anthropology in the UK.

Things really changed around the end of the 1990s, when suddenly forensics became sexy and you had forensic courses being set up in universities across the country. Suddenly people were becoming *teachers* in forensic anthropology, who'd never done a case in their lives and who were learning it one step ahead in the textbook. I have some sympathy with that because in the early stages I wasn't that much different. But within the last 10 to 15 years there has been a huge change in the professionalism of the discipline. And, of course, international and national judicial scrutiny is such that we have to know what we're doing – we can't play at it any more.

So did you actually set out to become a forensic anthropologist?

In my heart of hearts, I'm an anatomist. But the work just kept coming, and the big turning point for me was Kosovo. At that time I was working with Peter Vanezis, the forensic pathologist in Glasgow, and Peter was deployed with the British forensic team to Kosovo in 1999, very shortly after the Serbs retreated. He found himself faced with a crime scene that was an outhouse with 42 co-mingled bodies very badly decomposed: partly buried, partly burnt and partly gnawed by dogs. He said, 'I don't know how to do this, but I know somebody who does.' So at that point forensic anthropology became a subject within UK deployment.

And you were the first person to do it?

Yup . . . I went out to Kosovo about a week after the main team, and it was just, you know, 'How the heck do you do this?'

Had you ever had anything like that?

No, no. It had always been one or two little fragments at a time – a house fire, that sort of thing. But I was working alongside the Anti-Terrorist Branch, SO13, at the time, and these are *hugely* experienced officers. And with Peter Vanezis, who's a very experienced pathologist, we got through it all together, and sort of learnt it stage by stage. And Britain is very firmly entrenched within its forensic credentials, so absolutely everything was done to an evidential standard we knew would stand up to scrutiny.

Before we explore Kosovo further, I want to hear a bit about your early life. I understand your grandmother was particularly important . . .

That was my father's mother, and I spent a lot of time with her. She was one of these amazing ladies who could interact with a four-year-old as easily as she could with an 18- or 50-year-old. And she always had *time*, which I think is the most important thing any grandparent can have, because your own parents are so busy. She was the most adorable woman, she really was. She died when I was 15. She knew she was dying of lung cancer because she'd smoked a horrendous number of cigarettes throughout her life. She told me she was going to die, and I remember being very upset. But she did what is probably one of the cruellest things you can do: she said, 'But I'll never leave you! For the rest of your life I'll be sitting on your left shoulder to keep the devil away, and any time you need advice, you've just got to listen and you'll know what's the right thing to do.' [*laughing*]

And has she been there?

Oh, it's the bane of my life! There are so many times I've wanted to do something, and I find myself turning my head [*looking at her shoulder*] and thinking, 'No, she wouldn't be proud of me if I did that.' And I know that when the time comes, I'll actually have to face her, so I've got to get it right. What an *awful* burden to give your grandchildren! But she was a wise, pithy old lady and there was a huge hole when she died. She's seen me through all sorts of things, you know.

There have been some horrendously difficult times, but she's still there – 35 years later.

As a woman doing science, have you ever found yourself at a disadvantage?

Never. Any time that I've worked with the police . . . In Kosovo, for example – you're out there with a team of 18 men; you're the only woman; there's no toilet, so when you want a pee 18 men all have to look away. Never once has any one of them made me feel uncomfortable because I'm female. They are, in many ways, more protective.

One of the most disruptive things you can do for policemen is to have a young, blonde, available female on the team. You need a mother figure. They respect you for that, and they'll protect you.

So you've found that role has fallen to you?

Oh, absolutely. They want to talk about their families, about the things they see, and how they feel about it. They want to talk to someone who's not going to be a threat, and I think it's a huge compliment that they're prepared to do that, because policemen don't give personal information very easily. But they've never, in any negative way, treated me differently because I'm a woman. And coming to Dundee, being a woman hasn't made any difference whatsoever. If you can do the job and achieve the goals, then you're the same as everybody else.

Back to Kosovo – what were the challenges?

Well, the first challenge was the first site – those 42 men who had been herded into an outhouse. The gunmen had stood at the door and sprayed it with Kalashnikov fire. The chap who managed to get into the corner first was a survivor, and it's important for an indictment that you have one. The gunmen's accomplices stood at the windows, threw in straw, and torched the place, so that when we arrived all the bodies were huddled into one area because they'd been trying to get away. There had been six months of decomposition, so there was very little recognisable soft tissue left. 'Big, boiling masses of maggots' is the only way to describe it. Partly burnt, and dogs had gone in and taken away bits for food.

Our job was to document the evidence. If this is going to be an indictment site against Milošević, then the witness statement has to match up with the forensic evidence. If what the witness is saying is not borne out by the evidence, or vice versa, then that's not going to be a site that's likely to lead to conviction. So we literally had to start at the door of the building, on our knees, sifting fingertip through every piece of rubble. Once you got to what was a part of a body, or you could perhaps outline as a whole body, then you would lift it, take it away and do the post-mortem on what was left. Again, it's about establishing: is this male or female? How old? How tall? Is there any clothing or documentation? And literally working your way through that room until you've cleared it – bearing in mind that there might be explosive ordnance there as well.

I was going to say, what were the dangers?

We had an explosive device left for us at that scene. You have to make things funny – it's not at the expense of the deceased people, but it's what keeps you going in these difficult situations. There was a tree next to the crime scene, and we'd use that tree when we needed to take care of bodily functions. The first person to do that was one of the Anti-Terrorist Branch explosive ordnance officers. He came back beaming from ear to ear, and said, 'I've found a device!' They'd planted a device near the tree, with a trip wire so that when we walked down the path it would go off. And he was so delighted, first because he'd found something to do, but secondly because he'd actually been relieving himself on to the device and he was really impressed that he could still stop peeing in mid-flow at his age!

We had to blow up that device. And we had grenades that were placed underneath bodies with the pin removed, so that when you lifted the body the grenade would go off. You'd find razor blades, hypodermic needles in pockets, things that would inflict pain.

And how did you deal with witnessing such horror and such obvious cruelty?

Well, by that point I'd probably done 10 or 15 years of forensic work. I might not have seen anything on that scale, but it's the same principle. To work in forensics you need a clinical detachment, because you're there primarily to retrieve evidence. If you do become affected by it, you become inefficient in your objectivity. So you actually close down emotionally. Where it kind of breaks through is when you're really tired, when you're really hot – you know, when there are other stresses. Then sometimes it can boil over, but the majority of the time it doesn't.

I think anatomists learn this gradually as they're exposed to the human body. The first case I went to was a microlight pilot who went down off Inverbervie. He was a decapitated torso. I did that case with my supervisor. You become more and more able to cope with things that are difficult. But what you don't forget is that Post-Traumatic Stress Disorder can hit you any time. It may *never* hit you. It may be tomorrow, it may be a month, a year or even 10 years, so you have to be aware of it. But I can say, with my hand on my heart, I've never had the flashbacks or the lack of sleep . . .

In Kosovo did you have the relatives around, and does that unnerve you?

We didn't on that particular occasion, because we were so close to the time when the Serbs had retreated that the refugees hadn't started coming back. But they did very quickly after that, so we soon began to get onlookers. These were the relatives, neighbours, friends, and that adds an extra dimension, because the last thing you want is to add to their grief. So you take on board the responsibilities of your own job and the added responsibility of dealing with people who've gone through things we can only imagine.

Often it's very humbling. They felt they had to give you something, so they'd come with a cup of coffee, or cold water, and that was almost more difficult because they were thanking you for what you were doing.

But the one that will stick in my mind forever was a man who lost his entire family. A rocket-propelled grenade took out his trailer and on the trailer were his wife, his mother, his sister-in-law and their eight children. They were all literally blown apart. He retrieved as much as he could and buried them, which is a tremendously brave thing to do . . . To be able to go round and pick up what's left of your family, from an 18-month-old baby to your twin 14-year-old sons. And then we come along and say, 'Look, the UN has identified this as a potential indictment site; we'd like to exhume what you buried. Are you okay with it?'

We don't have to get consent, because a UN indictment would override that, but it's always best. He said, 'Yes,' and he said, 'What I want more than anything is 11 body bags. I need to put every one of my family into the ground with a name, because it's the only way God can find them. At the moment they're all together, and I need God to find my daughter; I need God to find my wife; I need God to find my mother . . .'

What we brought in filled a body bag and a half. That was all we could find, so there was huge pressure. I actually sent everybody out of the mortuary for the day and said, 'I've got to do this one myself,' because juvenile identification is my area of expertise. So I laid out 12 sheets along the floor of the mortuary and, going through the material I had, I started to separate little bits out . . . You know, 'That can't be the five-year-old, so is it the eight-year-old or the six-year-old?' By the time I'd done all that, we had something that I was absolutely happy represented each of the 11 people. But there were two 14-year-old boys, and all I had of them were arms, and they were pretty much bare bone by that point. I couldn't separate them because they were both male and the same age. One of them had a Mickey Mouse vest attached to it. I said to the policeman, 'Go and ask the father which of his children had a Mickey Mouse vest.' He came back with the name of one of the twins, and I thought: that's all we need.

We gave him back 12 body bags at the end of that day – the twelfth bag was what we couldn't separate. It would be very tempting just to split that between the bags on the principle 'He'll never know.' But that's not the point. She [*meaning her grandmother on her shoulder*] won't allow me to do that, because at the end of the day the man wants to be sure that in that bag is his wife; in that bag is his daughter . . .

It mattered very much to him, but it also matters to the courts because they could come along and say, 'Right, open up that body bag.' If what's in that bag doesn't equate to the named missing person you've said it is, then you're not a credible witness, and every bit of evidence you've recovered can be discounted. You can't afford to do that.

So we gave him back 12 body bags. And it was the most humbling experience of my life to hear him say, 'Thank you.' You think: God, for what he's been through this is the absolute least we could do.

So are there temptations to do a little bit extra for the family, or for the police, or are the limits of your responsibility very clear?

It is generally very clear. Most of the time we don't have involvement with the family, because you can't afford to be influenced by their emotion and their situation. So most of our work is in clinical isolation. And you go the full 110% on everything you do, whether it's for the police, the courts, the family or whoever, it doesn't matter. But when the family element comes in, then you do end up, I think, going that extra little bit that you possibly shouldn't. But you can't not.

In a place like that you were probably working very long hours with little sleep – how on earth do you look after yourself?

If you're the only anthro on the team and you haven't looked after yourself and they need an anthro, then you shut that team down. So there's a huge responsibility to look after yourself – to make sure that if you cut yourself you deal with it properly; if you get a tummy upset you deal with it properly, you drink enough water.

We also have a buddy system where you take responsibility for somebody else, who equally takes responsibility for you. If you start to see erratic behaviour, then you can pull them aside and say, 'We need to talk.' You'll do that for them, and they'll do that for you.

Also, part of the role of the senior officer in charge is welfare. We didn't have that in the early stages of Kosovo, so we did work far, far too many hours. It became clear that people were going to burn out quickly, and the senior officer said, 'No, today we're doing nothing. We're going to sleep late and eat well. You can read a book, phone home, do whatever you like, but we're not working today.' That becomes very important, but you can't do that in all circumstances – it depends on the nature of the deployment. If you're going in somewhere that's particularly dangerous you may have a very tight time schedule, and then you don't have the luxury of saying, 'We're not working today.'

What keeps you going?

It's the detective in all of us, isn't it? Our mystery is: who was this person? And when you solve that, it's a *huge* adrenaline rush. You think, 'Yup, someone's got their husband back. Someone's got their wife, their daughter . . . I've made a difference.' Even if it's not going to make a difference to the courts, it'll make a difference to somebody. And that's grand . . .

Working in a big team in Kosovo was a bit like Big Brother. You take a bunch of people that wouldn't normally choose each other; you throw them into a really difficult situation; you throw stresses at them – lack of sleep, lack of food – you make them work together. And yep, there are times when you will shout at somebody, lose your temper, but you know you've got to live with them again, so you get over it. It can be quite an experience! But the camaraderie you develop is *hugely* strong. And these are the people you can talk to, because, when nobody else will understand, they do.

The police brought some counsellors to Kosovo. And I mean, bless their hearts, they tried! But they never understood why we couldn't take them seriously. Because they'd never worked with us, they didn't know what we were doing.

So is this where the buddy thing came in – they suddenly realised you needed someone who knew the situation?

Yeah, absolutely. You'll sit down at night with a beer, and you'll just talk. But to have a counsellor come in and say, 'Tell me how you feel,' you think: for goodness' sake, I've been here for 12 weeks – how the hell d'you think I feel? It can be counterproductive if you haven't got the right counsellors.

In buddying each other, can you admit weakness?

Oh, absolutely. We had an officer out with us one time . . . We had been exhuming bodies in a field. It was miles from anywhere, so we couldn't take the bodies back to the mortuary; we had to do the post-mortems on a sheet laid out in the field. This was a group of women and children that had been massacred, and they were really in a dreadful state. We'd just exhumed the body of a little girl and she was still wearing her sleep suit and her little red wellies. One of the officers made a mistake – the little girl was about the same age as one of his own, and he put his daughter's face in his own mind on to this. I was working with the pathologist, and I looked up and thought: why's there a row of policemen looking at me? Then I saw that behind them was this officer who was falling apart: it was the men's way of giving him his moment of privacy and time to get over it.

So I took my gloves off, took my suit down and tied it round my waist, went over and gave him a huge hug. He broke apart, and he could then talk to me afterwards. We sat and we drank beer together that evening, and by the following morning he said, 'Och, I shouldn't have done that. I'll never do it again, because it's *not* my daughter.' I said,

'No, it's not,' you know? So you've just got to look out for it . . .

Going back to the beginning when you first did anatomy, how difficult did you find it? Were you ever squeamish apart from the rats?

When I was at school I had a Saturday job in a butcher's shop, so from the age of 13 I'd dealt with cold, red meat. I've never been squeamish about carcasses, or cutting up meat. It's natural. And to dissect a human body, to be able to look underneath the skin, is the most *fascinating* thing. It's a real privilege to be able to see what we're like inside. There's nothing ghoulish about it. We're all fascinated by bodies.

You were the first in your family to go to university, but how much support did your mum and dad give you over the years?

My mother was enormously proud. She had a scrapbook, bless her, of everything I'd ever done. My father is proud too, but he can never tell you. He's an ex-regimental sergeant major, classic Scotsman, and he finds demonstration of affection really difficult. But I know how much he cares.

How easy is it to take off your white coat, go home and be a mum at the end of the day?

Oh, easy. It's just the other side of the coin. There's only one occasion where I made a mistake, and I made a really big mistake. I have three daughters: 23, 12 and 10. I'd just come back from Iraq and had done a radio programme. I hadn't heard the final piece, but my husband had recorded it for me, and I went to listen. Grace, my middle child, said, 'Can I listen too, Mum?' I did a quick think, and said, 'That's fine.' I knew it was aiming for a middle-ground audience, so you're careful about what you say. But I'd forgotten that the interviewer had asked me, 'How do you reconcile the situations you can find yourself in and being a mum?' And I'd said, 'I believe I'm getting close to the point where I'm irresponsible, because my children need to know that their mother's coming home.' I looked at Grace, and her eyes were filling with tears, and she said, 'What d'you mean you might not come home, Mum?' I said, 'Well, I might miss the plane . . .' You could see what she was thinking: she wasn't fooled.

When I did the second tour in Iraq, she was the one who suffered. I phoned every night, but she suffered, and I thought: I cannot put her through that. So I turn things down. I won't go back to Iraq.

What about your husband? How supportive is he?

We've known each other since our teens, so we've been together a very long time. We went to university together, and he studied anatomy as well, but he learnt pretty quickly that if you want a decent income you don't become an academic! So he went into business.

But he understands what I do, and he's got a very good way of dealing with me. When I get home after being away he'll not question me at all. Three or four days later I'll start to tell him stories, and then he'll wheedle bits of information out of me. He's a very good listener, and we'll sit and talk about it, but it'll take a few days.

How would you say the extraordinary things you've experienced as a forensic anthropologist have affected your outlook on life?

The bottom line is that so many things don't seem important any more. I don't care if the floor doesn't get hoovered. It doesn't matter in the world if there's a scratch on the car. I care that my children get a great big hug every night before they go to bed, because you've just dealt with 20 children who will never have that again. So you go home and you hold your children tighter, there's no doubt. You value your family much, much more than you otherwise would, because you see the frailty of life. And that's what's important. The material side of things . . . it just doesn't matter.

WITNESS TO THE RAVAGES OF AIDS

Sebastian Lucas
Professor of Clinical Histopathology, St Thomas', King's and Guy's Hospitals, London

Sebastian Lucas grew up in a non-medical family in the south of England, and was persuaded to do medicine rather than animal physiology at Oxford University because it would give him more choices in life. The decision to specialise in pathology came soon after he began work as a medical student. 'I realised that the patients I was looking at on the wards were chronically sick and I just thought: "Can one face working with chronically ill people day after day? Is there a bit of medicine which is problem-solving, and where you can solve one case and move on to the next?" That's pathology.'

Lucas made his name in the early 1990s with his pioneering autopsy work on AIDS in Africa, where he investigated and underscored the powerful link between HIV and tuberculosis. The extent of TB surprised some doctors in the Western world, who believed TB was no longer a threat in the age of antibiotics, and had let down their guard. Having performed autopsies on more than 1,000 people who died of AIDS in Africa and England, Lucas knows as much as anyone about the multiple manifestations of the disease in different environments, and his findings have had a critical influence on the management and treatment of people with

HIV. However, as he freely admits, much of the vital research he carried out in the 1990s would be impossible in today's legislative climate. He got caught up in infectious diseases almost by accident, he says.

———•◦•———

I'll tell you about what got me into infectious diseases, and incidentally led me to where I am today. It was the drought summer of 1976 and I was at University College Hospital [UCH]. I was looking at a bowel biopsy late on a Friday afternoon when the rest of the department had gone home, and the case history said: 30 weeks pregnant, severe diarrhoea, query proctitis [inflammation of the rectum], query cause. It was a white woman, aged about 30. The usual diagnosis would be: nothing, or ulcerative colitis, or one of the standard diseases, but I looked at this biopsy and it wasn't any of these. It was inflamed, but it actually had amoebae in it. I'd never seen amoebae before, but I knew what they should look like because I was vaguely interested in infectious diseases even then. So I looked at a book and there was a picture of amoebae; I looked down the microscope and there they were, and I said to myself, 'This has to be amoebic colitis.' So I rang the houseman and said, 'This is amoebiasis.' What she answered was unprintable! Because, she said, the woman was in theatre at that moment having her colon taken out because it was perforated. 'She's got peritonitis, and she's probably not going to live.' And she didn't.

This was a Friday afternoon and the biopsy had probably been taken on Wednesday, because processing these things takes time. I was thinking, 'If we'd had the answer yesterday, might her life have been saved?' Well, you can conjecture till Kingdom come, I don't know. But the point is they thought she had ulcerative colitis and put her on steroids. That's what you do if it's ulcerative colitis or Crohn's disease, but if you've got amoebiasis it's the *worst* thing you can do: the disease just lets rip, perforates the bowel and you die. This is perfectly well known, but you have to think of amoebiasis first. And what on earth is a pregnant 30-year-old who's never left North London doing having amoebiasis, a tropical disease?

Not many people had ever seen a case of amoebiasis in this country. Certainly I never had. And certainly the main surgeon involved in this process never had. I remember presenting the case at the next gut meeting we had, and saying, 'Actually, this is amoebiasis,' and watching the physicians sink their heads in their hands and whisper, 'Oh, God.' Because this was a treatable disease . . .

So I did a bit of further investigation . . . As I say, this is what got me into infectious diseases: this actually changed my life. I found out, probably from the widower, that what had happened was that his wife hadn't felt like cooking and so the neighbours cooked something for her and passed it over the fence. Very nice of them. This is in north-west London, and they had just come in from Bombay. So basically the infection was on their hands. A goodwill gesture, sadly carrying rather unfortunate consequences. Because there is no endemic amoebiasis in that part of London.

And it went further, the bizarreness of this story. In those days there was an amoebiasis research unit in the Hospital for Tropical Diseases, which was run by an extraordinary guy – he was an air vice-marshall, one of the older Air Force medics. I gave him a potted version of this story, and said, 'There's no question about the cause of death: it was amoebiasis. As far as I can see, the infection came from the neighbours who had recently arrived from India,' where amoebiasis is endemic, and they were obviously symptomless carriers. I'll never forget his retort: 'This doesn't happen in the United Kingdom!' and he put the phone down. So the second message of this story – and I tell all the medical students – is: 'Don't always believe what old experts say!'

And infectious diseases are intellectually very teasing and exciting, are they?

Absolutely, yes. One of the things I do today, apart from looking at challenging autopsy cases, is a lot of infectious disease consultation work. People send me slides and say, 'What is this? I think it might be this,' or 'I haven't the faintest idea what it is.' Every week will bring in one or two real gems.

Two days ago someone sent me a liver biopsy, saying, 'What on earth is this?' I looked at it and thought, 'I can't believe this! I've never seen it before, but I know what it is.' It turned out to be someone who had had some abdominal trauma which had obviously ruptured a hydatid cyst, which is a parasitic cyst, and bits of that had gone into the liver circulation, blocked it off and damaged the liver. I didn't know that could happen, actually. I'd written about hydatid cysts in the standard textbook of liver pathology, but I'd never seen this version of it.

That's the thrill of this work: I'm looking at things which are either new versions of diseases I know, or genuinely really rare things, or even things that no one else has ever seen before. You're kind of working on the edge all the time.

Tell me about your involvement with HIV and AIDS.

To my slight shame, I didn't pick up on AIDS at the very beginning. It started nominally in 1981 when gay men in America started presenting with diseases no one had ever seen with any frequency before, if *at all* – that is, Kaposi's sarcoma and *Pneumocystis* pneumonia [PCP]. In the AIDS calendar, 7 and 14 July 1981 are big days. That's when Jim Curran of the US Centers for Disease Control [CDC] said, 'I think we have a problem.' He became the great man who drove the response to AIDS in the beginning. This is all in retrospect – I didn't pick up on that at all at the time.

In 1982, I was here at St Thomas' as a research fellow working on tropical infections with a charismatic pathologist, Michael Hutt, and we had a frozen section from a patient who had a lung problem, and no one knew what it was. Frozen sections are done during operations when surgeons want answers *now* so they can decide what to do. 'Is it cancer? Do we chop out the rest of the lung? Is there something else?' They're sitting there waiting for answers and it usually takes five or ten minutes. The patient was called Terrence Higgins [one of the first people known to die of an AIDS-related illness in the UK]. Yes, him . . . Quite!

And he was under the knife?

He was. Maybe they thought he had cancer, I don't know. Anyway, we looked at this and we just did not know what it was. But we knew it wasn't cancer, so they closed him up; no more procedures were done. We looked it up in the textbooks and realised it was *Pneumocystis* pneumonia, which was the first case most of us had seen.

At that point the penny dropped: this was what we were then calling GRIDS – Gay Related Immune Deficiency Syndrome. AIDS hadn't been officially labelled as that at that point. I remember that patient, Terrence Higgins, was looked after in a ward more or less to himself; he was treated like the proverbial leper. That's one reason why the Terrence Higgins Trust was set up – to say that people shouldn't be treated like this, and quite right. He died later of PCP and cerebral toxoplasmosis, a parasitic infection.

I left St Thomas' shortly after that to go to UCH, so I wasn't part of what happened here. But the name changed to AIDS, and then a group of people interested in AIDS grew up here at St Thomas'. Anyway, I forgot about it.

I was at UCH from 1983 until 1995, minus a year in Côte d'Ivoire in 1991/2. I got interested in AIDS not through anything going on at UCH, but actually because of Michael Hutt. He's the grand old man, the main person in my professional life. Back in 1977, when I first realised that infectious diseases were interesting, I said, 'Who knows about infectious diseases in Britain?' and everyone said, 'Michael Hutt.' He was the professor of geographical pathology here – a unit set up to promulgate the importance of tropical pathology.

Michael Hutt had been at St Thomas' in the 1950s as a junior doctor and a trainee pathologist, and then as a consultant pathologist. Then he decided to do something different, and he went to Makerere Medical School in Uganda as the pathologist. He was there through the 1960s. Mike and people like him – and there were a lot, all very famous clinicians, epidemiologists – made Makerere a jewel. There were loads of exciting things happening; it was virgin territory for describing diseases.

Mike Hutt retired from St Thomas' in early 1983, and he gave me his practice, which was looking after diagnostic histopathology for mission hospitals abroad, particularly in Africa. I simply took over that work *en bloc* and built it up. It was partly cancer, partly TB, partly infectious diseases, partly just general stuff. These cases would come in by post. Surgeons in mission hospitals all over the place – but particularly East and Central Africa – would send in these specimens. They'd do operations, put the specimens in formalin, let them fix, then pack them up – usually in the ends of surgical gloves, which are very good containers, water-tight, tie them off with a piece of cotton or simply tie a knot in them like a balloon – and put them in an envelope.

We'd look at the specimens, write our reports and then send them back by airmail. I built that up because it was seen to be useful, and it was great fun. I saw loads of things I had never seen before – wonderful things, pathologically, that don't happen in Britain.

One of the series of materials we were getting from overseas came from the Makerere School of Medicine in Kampala, Uganda – as I have mentioned, *the* best medical school in East Africa then, as now really. Among the stuff was a set of intestinal biopsies taken from people in Mulago Hospital under the auspices of two very active and dedicated clinicians. One was Nelson Sewankambo, a great Ugandan friend who is now Dean of Medicine at Makerere, and the other was Rick Goodgame, an American Baptist missionary doctor, who was an enormous enthusiast. They sent me all these bowel biopsies from people with slim disease. 'Slim' was the name given to a condition being seen in the early 1980s, and which we now know was AIDS. At that time, of course, we didn't have the virus; HIV wasn't known about till 1983. This condition had started in the south-west of Uganda, and had moved up country. The patients had wasting, and they had terrible diarrhoea, like cholera, and they wanted to know why. So the doctors started doing biopsies but the local pathology lab couldn't cope, so they sent them to us.

It was obvious that there were some very strange things going on. One of them was a huge amount of an infection called cryptosporidiosis, a gut parasite that you don't see much of now, but it was big then. It became evident that slim disease was gut cryptosporidiosis. We had a series of about 23 cases and we wrote up our findings in a paper with the title: 'Ugandan AIDS: wasting disease is cryptosporidiosis' – something like that. It was the first article published in a new journal called *AIDS*. Volume one, page one was us! Goodgame, Sewankambo, and my name tucked on the end. And that got me into the AIDS scene.

Then, as that was brewing up, two things happened. One was that the First International Conference on AIDS in Africa was held, organised by another remarkable clinician called Nathan Clumeck, a Belgian, who was working in Rwanda, where AIDS was also big. At that time, 1983/4, the first accounts of AIDS in Africa being an issue came out from Projet SIDA, which was a US-sponsored project based in Kinshasa and run by great people like Jonathan Mann [who died in an air crash in 1998], Peter Piot, Marie Laga, Anne Nelson and Robert Ryder . . . All these names! All history now. They were in Kinshasa in 1983 and they published a big paper in the *New England Journal of Medicine* which said: 'There is AIDS in Africa, and it's the same disease as in America, it just looks different.'

At the same time, Nathan Clumeck published a very similar paper which said, 'We're seeing the same thing in Kigali as well, and in Burundi . . .' So gradually these reports were coming in, and Nathan organised a two-day conference in Brussels in November 1985, which was attended by about 100 people. At the end there were two press conferences held, one organised by Nathan that said there was a big problem of AIDS in Africa, and the other by a few black African physicians that said, 'There isn't a problem'!

Anyway, I met several important people at that conference, including Kevin de Cock, who is the most important AIDS physician–epidemiologist around now and is my best medical friend, really. After the conference we wrote a document for the Dean of the London School of

Hygiene, saying that the School should get into AIDS in Africa 'because AIDS is going to be very, very big'. In fact it is much bigger than we ever thought it was going to be.

Kevin then went back to CDC in Atlanta, and we remained in touch. And then I got a phone call from an extraordinary surgeon working in Kampala called Wilson Carswell, who said, 'Come out. We know you looked through all this gut stuff and we've got to document it properly – AIDS is ravaging this country, it's dreadful; come and see the hospital, you won't believe it. Come and stay for a fortnight.' Wilson met me at the airport; took me home for a shower and then to the hospital and just said: 'Look.'

I have never seen anything like it. These are enormous wards, and every bed was occupied by a dying skeleton. I was staggered. Again, it flicked a switch: this is interesting, this is worth doing, this is a disaster. Wilson said, 'It wasn't like this 10 years ago. Something's happened, and *you* are going to tell us what.' I said, 'Well, it's AIDS.' And he said, 'I know that, but what have they got that's making them like this? We know some of them have *Cryptosporidium*, but what else?' And I said, 'How about TB?'

I looked at patients, spoke to the pathologist there, did a few autopsies myself, took photographs, and we began to feel that, actually, 'an awful lot of what we think is HIV-related disease may well be tuberculosis – made worse by HIV'. Which turned out to be entirely true.

When you saw all those emaciated people in the wards, had they been diagnosed with TB?

No, they hadn't. It was a concept that gradually built up and finally became completely solid in 1991. But that's five years later. So, I grabbed material – with consent – and brought it back to the UK.

I'll tell you a good story. On my last day in Kampala, the British Council gave me a Landrover. January/February 1986, Kampala; not a good time to be there. Museveni had only just conquered the capital, and was moving north to try to get rid of the rebels from Gulu and beyond. Getting around Kampala was bloody difficult – pot holes, shell holes and what have you. Nelson Sewankambo and his great buddy David Serwadda – another Ugandan public health physician who's become very famous in AIDS – said, 'We should do a quick study. Is the virus (because it was still not HIV, it was just 'the virus') transmitted horizontally?' In other words, if you live in the same household as someone, but don't sleep with them, do you get it? 'How can we answer this? What we need is a lot of blood samples very quick. Let's go down to Rakai District, where this is supposed to have started.'

Nelson said, 'I know the district medical officer of health; I'll ask which are the worst-affected villages and do we have his consent to bleed everyone and bring the samples back for analysis.' The DMO said, 'Yes.' So we drove down there in the Landrover at dawn. The DMO had a map and he said, 'Go to these villages here.' So we went with our bundles of syringes and needles; we found the village head man and asked, 'Do you approve?' (This is what consent was in those days!) We bled about 100 people in two or three villages. At the end of a long day we came back with lots of little vials of blood, all labelled and coded and so on. My task was to get these to England, and to the public health laboratory in Porton Down – and in particular to Bob Downing or Graham Lloyd, two more famous names in HIV virology, who would do the analysis.

The next day I left Kampala, having spent two extraordinary weeks that changed my life. I had a Thermos flask packed with ice and containing 100 vials, and in my suitcase I had loads of tissue blocks of AIDS pathology.

I kept thinking, as I passed through several airport departure stations, 'If someone asks me what's in this flask, what am I going to say?' But no one ever asked! A different world now, isn't it?

When I got home, a courier took them away to Porton Down. It became a paper: 'The AIDS virus is not transmitted horizontally.'

As I say, we were also wondering about TB. While I was still in Uganda I thought, 'HIV makes TB worse in many contexts, but how can we prove this? I wonder if the pathology looks different?' We obtained tuberculous lymph node biopsies from patients with AIDS, analysed them and published our findings. It was the first description of tuberculosis pathology being very different from standard TB when you've got HIV disease. The bacterial loads in these people were just colossal. If you do the stain for tubercle bacilli, which we call a Ziehl–Neelsen test, the bacilli come out red. With these cases you didn't need to look under the microscope – if you just held them up to the light the whole slide looked red! Phenomenal densities.

However, the sheer overwhelming importance of tuberculosis didn't really sink in until the early nineties – another half decade. And that happened in part, I like to think, from work we did in Côte d'Ivoire.

How did you find yourself working in Côte d'Ivoire?

Kevin de Cock had started a project in West Africa. He was asked by CDC to investigate HIV-2, which had just become evident. CDC didn't want to be caught with its pants down a second time, because they had kind of misfired with working out what HIV-1 was. (In fact, *no one* did very well – Heavens, this was a new disease!) So the CDC said, 'We need an HIV-2 project; go and find the place to do it.' It became evident that the only place he could possibly work was Abidjan, because the communications in alternative places didn't work. And there was a big American Embassy there, so the Retrovirus Project, Côte d'Ivoire – a collaboration between the CDC and the Ministry of Health of Côte d'Ivoire – was set up with American money. A new building went up in the grounds of the huge hospital; Kevin set up his unit and they started doing basic HIV work, with HIV-2 as the added interest.

Sometime around 1989 he said, 'Come out to Côte d'Ivoire to see what we're doing and see if it interests you.' So I went out for two weeks in early 1990. I was appalled by the climate, and I was staggered by the amount of work going on. They had a wonderful lab. Most of the scientists were Ivorian – basically, they'd employed the best of the medical and paramedical Ivorians to work there, and paid them salaries which were somewhere between Ivorian and American. And they had very good infrastructure, with computers the like of which I'd never seen before. I met a lot of doctors. I met the pathologist, who was very nice. I looked at the mortuary, and thought, 'God almighty, what a ghastly place.' They all said, 'Come and work here,' and I said, 'No, no, no!'

But then Kevin said, 'Would you like to finish off a project for us?' He gave me a whole lot of tissue blocks and slides. They turned out to be a set of autopsies, and as I worked through them I found that they were 50/50 HIV-positive and HIV-negative people. The HIV-positive people had TB and a bit of *Pneumocyctis*, and the HIV-negative people had boring things like lung cancer. And when I looked more closely I saw they were, in fact, consecutive deaths – every person over a period of, say, three months had been autopsied. I thought, 'You can't do that in England!' I asked Kevin, 'How on earth did you get consecutive deaths autopsied?' And he said, 'Ivorian Law, Napoleonic Code, it's French. Anyone who dies in a teaching hospital can be autopsied without consent required.' I said, 'Kevin, I'm coming back!' And he said, 'I knew you'd say that. That's why I got you out here.'

I still have nightmares just thinking about 1990 and the panic to get things going. We got a project that seemed okay but it would never pass muster now – ethically, no one would even *contemplate* doing this today. We drew up a project to autopsy as many people with HIV – and some HIV-negative controls, including children – as we could in a year. God, it was hard work! My own kids were doing O and A levels, so I went there by myself, and just worked flat out essentially for a year. But when people accuse me of being bad – which they do, very occasionally – I say, 'They wanted it. The Côte d'Ivoire Ministry of Health supported this to the hilt, and more! And that's what was done then.'

What manifestations of AIDS did you find in Africa that were different from in the West?

I was doing a lot of HIV work in London by then. UCH and Middlesex Hospital refurbished a mortuary dedicated to HIV for me in 1989, and we did loads of consented autopsy cases, and a few coronial ones. They all seemed to be the same then: all *Pneumocystis*. All the patients I was looking at were gay, white, middle-class men – that's what AIDS was in London then. Africa was completely different.

The first thing that struck us in Côte d'Ivoire was, 'Actually, it's all TB.' More than half of the cadavers I looked at had tuberculosis. I couldn't believe it. This became evident after about the first two or three months, and Kevin and I said, 'Well, even if the project comes to a stop now, we've proved what it is, it's TB.' So you'll find, in the early nineties, loads of papers written by Kevin and me and others just banging the TB drum, and stressing the importance of diagnosis, prophylaxis, etc., because we *knew* TB was the major problem.

I should say that whilst we were doing this project, the clinicians in the hospital in Côte d'Ivoire rapidly got wind of it and every Saturday morning there was a clinico-pathological meeting of the cases we'd seen that week – with the professor of that particular ward, his senior doctors and all the junior doctors as well – and they all loved it. No one said, 'Ethically you shouldn't be doing this.' They all said, 'This is fantastic. We've never had feedback like this; this is *hot* feedback. The patient died on Monday; we now know why he died. This has never happened with us before.' They gave me every possible help. This was real clinical pathology going hand-in-hand with management of cases. They gave me a huge party when I left, and said, '*Please* come back! This is what medicine's meant to be all about.'

I came back to the UK in 1992. And the next two or three years were basically spent writing everything up. It doesn't happen overnight; there's a lot of analysis. Kevin was also back in Britain, which was wonderful. He was a professor at the London School of Hygiene, so we sat together and wrote papers. And that enabled me to get the chair here.

Looking back on your early research into AIDS and how easy it was to collect material, do you feel the Alder Hey debacle was waiting to happen, and that it was right that people should start challenging the practices of pathologists?

Only up to a point. And that's because, when autopsies are done well, the public health benefits are so colossal that in my opinion they override, to some extent, personal objections. I'd be quoted saying that in public quite happily because it's true, but it doesn't look good. I had two complaints throughout the whole process. No parents of any of the kids in Abidjan complained, and we did about 150 children. So my point about overriding what we now regard as narrow consensual requirements is that it's completely outweighed by the public good it does.

When Alder Hey happened I was here at St Thomas', and we were more vulnerable than most other places in the country because we have the biggest collection of paediatric kidneys, hearts anywhere – much bigger than Bristol, because we're a big cardiac surgery unit. We had a professional perinatal pathologist who was very interested in congenital heart disease, and she kept everything, because one did – no criticism of her. She took the whole Alder Hey and Bristol business very badly, and after some months she left. Now the last thing she'll ever do is perinates – all that phase of life, all gone; she's doing other things now.

I have to say the chief exec here, Sir Jonathan Michael, was brilliant. I have only praise for him. He realised that organ retention was a big issue, we were vulnerable and it had to be addressed openly. So very rapidly we had a team organised to work out what we'd got; that was the first consideration and there was a lot of work to do.

We have the biggest pathology museum in the country here, the Gordon Museum, in Guy's. Not for a moment did we shut a door, remove an exhibit, close anything. Unlike Dublin, which closed its museum. Unlike the London Hospital, which cemented it in! I joke not. A whole lot of deans took fright, and actually shut the museums. We said, 'We do not do this. We will keep our displays open.' And particularly, we have a fantastic series of pots of malformed fetuses. Now this is *not* prurient stuff; this reminds the students that this is why we have prenatal scanning: so that we don't have any more babies born with no heads, with spina bifida and things like this. It's a brilliant display. You can be lectured like mad about the importance of antenatal screening; 10 minutes in the Gordon Museum and you can see why we do it.

One thing that didn't happen here, and that *did* happen in Bristol, Alder Hey and other places, was that we didn't have a whole lot of anguished relatives on our doorstep. I think the reason is that London is very different from Liverpool: patients coming to this hospital come from an enormous catchment area, so big that they never get together. So from that point of view the reaction was diluted. We talked to people, and we were not sued. We handed back some material, but most people were not interested.

But looking at Alder Hey, was what Van Velzen did standard practice for the time, or was he actually behaving unethically?

The major criticism of what Van Velzen was doing was that he took much more than anyone else did. Secondly, a very legitimate criticism is that he actually lied, in that there were reports coming out that said he'd done work on these materials which he obviously hadn't done because they were still intact. That was professionally unforgivable. He took stuff intending to work on it but never actually did – and probably knew he would never be able to because he'd just got so much material, but he just kept on collecting.

But does that not go into an archive that other people can use?

Yes, they could do if it was properly documented. But it wasn't properly documented, so that archive is effectively useless to anybody else. And he didn't tell people he was doing it. Now there's a lot of discussion as to whether he needed to have told, in all circumstances, what he was doing, but I'm not going to revisit all that, because it was a mess.

Actually, things haven't got much better. They've got more bureaucratic, but there's still a huge grey area in tissue retention. It's very clear what to do if you want to be squeaky clean: you do nothing. That's easy. But the point is, to be good and to be useful for public health you need to do a bit more than that, and that's where the grey areas come in. I experience this day by day, and I know how to deal with it because I've rubbed along the raw edges; I've worked out how to do things that are ethically reasonable with appropriate consent built in, but so that you get what you want in order to advance knowledge and help people.

So are the rules now so constraining that you're actually losing an awful lot of what might be beneficial?

Absolutely.

So has that in any way dimmed your enthusiasm for the work? What drives you to continue?

Curiosity! Even though I'm 60 I'm still very, very curious about things, and I haven't yet seen it all. It's a real important driver, actually, curiosity. It's probably the biggest driver of all. I come into work every day early and leave later than my wife would like because there's such an amazing amount of things to see, and some of it I've never seen before. In a way, every case is a challenge – you think: it's me against the truth, and can I find it?

HOW BODIES DECAY: LESSONS FROM THE BODY FARM

Bill Bass

Emeritus Professor of Forensic Anthropology, University of Tennessee, Knoxville

Around half of all forensic anthropologists working in the US today have studied with Bill Bass, who, in the early 1970s, set up the Anthropological Research Facility – better known as the Body Farm – on waste ground at the University of Tennessee, at Knoxville. Bass, who started his career excavating burial sites and cataloguing the bones of Native Americans for the Smithsonian Institution in Washington, DC, discovered from experience that one of the hardest things to get right is the length of time a body has been dead. Working later as a consultant for the Kansas Bureau of Investigation, he became convinced there was a need to study, in a methodical, scientific way, exactly what happens to bodies as they decompose. But it was the case in which his judgement of the time of death was a full 112 years out that galvanised him finally to set up the Body Farm, where corpses are left to rot under different conditions, and all manner of studies are undertaken by scientists.

When I went to speak to him in autumn 2007, he drove me straight from the airport to the site on a hillside overlooking the Tennessee River.

Let me give you just about a three-minute history. I taught forensic anthropology for 11 years, from 1960, at the University of Kansas. I also identified skeletal material for law enforcement agencies in Kansas, but I don't ever remember getting a maggot-covered body – they were all skeletal remains. I came here to Knoxville 1 June 1971 to take over a three-person department that was an undergraduate-only programme and to build it into a graduate programme. The medical examiner in Tennessee knew me and asked if I would serve as a forensic anthropologist for the medical examiner system, and I said, 'Yes.'

It wasn't long before bodies started coming in. Now, the police don't ask you, 'Who is that?' They ask you, 'How long have they been there?' I think the reason is that in a criminal justice system they're trained that the sooner you get on the chase the more likely you are to solve a crime.

The major characters in body decomposition are the maggots – I mean, the flies are the first ones to get there, and they do the greatest reduction of a body. Well, I didn't know anything about maggots, so I looked in the literature and there was very little in there. I won't say there was nothing, but there wasn't much dealing with length of time since death. So I decided, 'You know, if I'm talking to the police about how long somebody's been dead, I better know something about it.' So I went to the dean, and asked if I could have some land to put dead bodies on.

I started up with the sow barn at the Holston Farm, which is about 12 miles up the river from where we are right now. We used that from 1971 until 1980. Now business was really expanding, so I asked for more land. Where we are right now was where the university used to burn its trash, and sometime in the 1960s the Environmental Protection Agency said, 'You can't have open burning.' So they covered this over with dirt and it just grew up with bushes. And I reckon they figured, 'It's been a dump for all these years, might as well go give it to Dr Bass for his dead bodies!' There are somewhere between two and three acres right here.

This is way over-used, by the way: there are about 150 bodies out here right now. And we don't have any sterile land, if you want to call it that, for a student wanting to do research – there have been bodies put almost anywhere you can put a body. So we went back to the university and got some additional land.

Bill, there was a story I read somewhere about what made you think of doing this. As you say, the maggots, but wasn't there something also about being asked to judge how long a body had been dead and you miscalculated by . . .

One hundred and twelve years! [*we laugh*] This is the Colonel Shy case. One of the well-to-do families in Tennessee at the time of the Civil War in the United States was the Shy family, and one of the sons, William, became a colonel in the Confederate Army. He was killed in the Battle of Nashville in 1864, within about 15 miles of his home, and he was taken back and buried in the family cemetery.

As time goes by, the Shy family moves away. The house had been sold a couple of times and it was being remodelled. The wife of a physician had bought it, and she was checking on the remodelling and found out there had been a grave disturbed in the family cemetery behind the house. She called the sheriff and he came out and decided they needed me. So I went over, and we got the body up. There was no skull, but the body was so well preserved that you could see pink tissue on the femur. And I thought – in my frame of reference as to how long somebody would be dead and still have pink tissue on them – that they probably hadn't been dead more than just a few months. So I thought: you know, one of the ways of getting rid of a body is to go to the cemetery and dig a hole on top of a grave and put the body in there. Well, that was on a Friday. By Monday we figured no, the person we had was really the Civil War colonel.

What had happened was that when Colonel Shy was killed, he was embalmed with arsenic and was buried in a cast-iron coffin that did not leak. Nobody had ever looked at what happens to bodies in coffins; nobody had ever looked at arsenic as an embalming agent. And so this was the beginning of a whole series of studies.

So Colonel Shy was the clincher – the case that convinced you of the need for a facility where you could study decomposition?

Exactly. That was the straw that broke the camel's back, yup. All right, so let's go and look. [*Bill unlocks the gate*]

How much security do you need in a place like this?

A lot. My people aren't going to get out, but everybody wants to get in to see my people! We put the chain-link fence up first and found out that doesn't do much, so there's razor wire around the whole thing. And there's a camera trained on it at night to make sure there's nobody out here.

Can you get a little smell there? You will in just a minute. Although this is a good time to come, because you don't get much smell in the winter. Summer time is bad, oh yeah!

Now, how about if I take you on a 15-minute tour, just to give you an idea of what goes on?

See that little 'Bobcat' there? [*pointing at a mini-forklift truck*] When I first started teaching, all of the graduate students were males, and then we began to get a few females – and now every one of the graduate students we have is female. We have five graduate assistants assigned out here to make this place work. Now we are a population in America that's obese, and you can't get a 90lb girl to pick up somebody that's 250–300lb, so we now have a Bobcat to pick up bodies and carry them around.

[*We walk around, our feet shuffling through the leaf litter*]

This may be a little muddy . . . Now, this is a concrete slab right here somewhere. [*pushing leaves off with his foot*] We in America tend to kill our husbands and wives off, and if you kill your wife, what do you do with her? Well, you can take her out in the yard and bury her. But then the neighbours look over and say, 'What are you digging?' So you dig a hole to put your dead wife in and you cover it with a concrete slab so you have a little patio area where you can sit out.

Now, it's very difficult to find bodies like this. The best procedure right now is to get what's called a 'cadaver dog' – they come out and sniff around the edge and pick up the smell of decay. But what we're looking at here was a Master's thesis using ground-penetrating radar – it goes through the concrete to pick up the bodies underneath. We had six of these slabs; this is the only one left. There were bodies buried at different depths and we looked at those over a period to see if you could use this technology to find buried bodies.

And did it work?

Yeah, it worked. We covered half the bodies with trash and junk, and the other half with dirt and rocks. But you could see what was going on under there.

You'll see that most of the bodies out here are covered with black plastic – see a skull over there? Maggots do not like sunlight, so if you have a body uncovered the maggots will leave the skin as an umbrella and they'll eat all of the interior organs away. It looks like the body's in pretty good condition, but it's not – I mean when you get there there's nothing but the skin, like leather, over it. What we're doing is trying to get the skeletons down to nothing but bones, so we can study the skeletal material, so that's why we put black plastic on most of them.

How long do you generally have a body out here to study, and then what do you do with it?

Normally not more than a year, and then we take them in and clean them up. Each of them has a number. For example, the first body in 2007 is 1/07. When they go in to be cleaned up they end up in a box 1ft by 1ft by 3ft, and that number stays with them forever. We don't have the names assigned to the bones or the boxes; you can look that up on the master list.

And what are you looking at other than the maggots and the ground-penetrating radar?

Okay. The decay of the soft tissue: how long does it take? How long does it take for the right arm to fall off? What happens to the hair? What happens to the fingernails? I mean, you name it, there's a graduate student either has done it or is going to do it one of these days. Most of the research out here is either a Master's thesis or a doctoral dissertation.

I'm going to show you a project now. You see this pipe sticking out of the ground? Remember I told you earlier that we bring the dogs in to see if they can smell a decaying body? As we stand here today, nobody has ever found out what cadaver dogs smell. So we have buried four bodies and there are these pipes that run down and through the body so that we can catch the compounds that are given off of decaying bodies without disturbing the body. We just unscrew the cap, suck a little air out to get the compounds, and put the cap back on.

We have found over 400 compounds. The body as it decays gives off what are known as volatile fatty acids – 'volatile' meaning that they dissipate. I'm going to show you volatile fatty acid stains on the ground. They're liquid, and they will kill the vegetation right around the body as the fluid leaches.

Now, obviously not all of those 400 compounds are important, but there would be probably 10, 15 or 20 that are. We're at the stage now where we've been able to chemically isolate these compounds, and one of my graduate students, Arpad Vass, has just designed a sniffer to try to replace the dogs. We're trying to do two things. Once we find out what the major compounds are we want to bring the dogs in and say, 'Okay, d'you smell A, B or C?' And then you can take Arpad's hand-held sniffer, walk across the ground and it will indicate when it picks up some of these compounds. Eventually we'll have something that will go on a vehicle: you can drive across a field and it would tell you whether there are buried bodies out there or not.

Are there differences in the decaying process in different climatic conditions?

Oh yes; the major factor in decay is temperature. You decay faster in the summer than you do in winter; you decay faster in Florida than you do in Wisconsin.

You see the white material right there? Well, that's adipocere – that body decayed when it was wet, and the fat has turned to a soapy-like substance known as adipocere – it's called 'grave wax' in the old literature. I'll tell you what they look like, the really good ones: they look like they do in a wax museum. I had a woman killed her husband and she 'buried' him in the basement. It was a wet basement and he was dead about five years when we found him. Why, you could look at him and tell who he was, no problem at all. Every whisker was in place . . .

Why hadn't the maggots got at him?

Oh, because he was in the basement of a house, wrapped up in plastic. The maggots don't get to you if you're buried, essentially, because the flies don't dig down to get at you. They may dig a little bit, but not much.

Look at this group of bodies right here. There's your skull. Now that dark stuff right in the middle there, that kind of goo, that's volatile fatty acids. This one here has dried off. This is mummification, where the soft tissue has simply dried up and dehydrated. It was probably open to the sun and wasn't covered up . . .

And under those conditions would the maggots get inside and take the organs out?

Yes. You see those holes in the tissue? Those are where the maggots have come out from inside.

Where do you get all these bodies from?

Well, we get bodies from three sources. We get unclaimed bodies that come through the medical examiner's system. These are people who have been killed, or for some reason ended up in the medical examiner's system to be identified. And if they're not claimed, the cost of the burial falls on the county in which the death occurred. It costs about $700 to bury a person, and they'd really rather give me the body for nothing than pay the $700.

Second, you have a husband or wife decide, 'When I die I want my body donated to science,' but they never do anything about it. Then one of them will die, the mortician will come to pick up the body, and the surviving spouse will say, 'You know, they always wanted their body donated to science.' So the mortician will call me. And now I've been on television enough that we have over 1,000 people who have willed us their bodies. Up until 2003 most of the bodies we got came from the medical examiner's system. Since 2003 the willed bodies have been in the majority.

We have a yearly memorial service, in which we randomly pick one skeleton for burial and that individual represents everybody in the collection. A few families come and take part in the little service that we have.

Here's one body, by the way, that was not covered and you can see how the skin has turned to leather.

The hair is still pretty good, isn't it?

Hair will last quite a while. I have excavated burials that were 100 years old and the hair was in good condition. Hair and fingernails tend to hold up. Bones and teeth are the best – they will last, but you don't want them to get wet. All right, let's just wander on up the hill.

It seems amazing that you were the first person to think of doing something like this. There must have been masses of people frustrated by the lack of information about how bodies decompose.

Obviously people had been interested in it. But medical examiners learn on the job, so it was more 'anecdotal' information. And when that medical examiner left or died, all that information disappeared, you see. I never set out to do anything that was going to be world famous. One thing just led to another . . .

Do you not find animals getting in and disturbing the bodies?

The fence keeps out the dogs and the coyotes and things like that. It does not keep out racoons – they're smarter than people. They climb trees, and they drop down in here. That's what that camera is for, right there. We've been doing a study on night predators. There are squirrels, rats, skunks and racoons, and they will all eat on bodies. They're one of nature's ways of reducing a dead body, and we have published a little in this area. We've been trying to look at what the scavengers do to bodies.

[*We look across the Tennessee River at the university campus, as we start to leave the facility*]

What do you personally feel about these bodies? Are they research material, or are they former people?

Okay, that's an interesting question. I have lost two wives to cancer. I hate death. I hate mourning. I hate funerals. I just don't like any of that culture at all. But I never see a forensic case as a dead body. I see it as a challenge: do I have enough knowledge to figure out who that individual is and what happened to them?

[*We drive to his office in the dingy old changing rooms of the giant sports stadium. He takes me into the storeroom with all the boxes of bones*]

These are all forensic cases . . . Here's a good one, hear that? [*rattling*] That's his brain. Now, you see the eyebrow, see the hair? You can tell that this individual has been embalmed. The skin looks like old paint, where it's kind of dried up.

There's one story after another in here. Let's see what this one is. Okay, you see the spray painting on the top? This is a Japanese trophy skull brought back from the Second

World War. This was a Japanese fighter pilot from Okinawa, and the guy that found him and brought the skull back was from East Tennessee.

Did a lot of people collect Japanese trophy skulls?

Yes, I have two in this collection . . . Let me tell you the story about this one. We have a lake west of us, and one time a fisherman hooked on to a body that didn't have any head. They brought me the body, and I said, 'Gee, we really need the skull.' This is a rural community with a weekly newspaper, so it was a big story. You know: 'Fisherman finds dead body . . . police looking for skull.' In the next two days, three skulls came into the police department and this was one.

Now this skull came in from a man who buys junk, old cars and things. It was sitting in the motor compartment of an old car he'd bought. Couple of weeks go by, then he reads: 'police looking for skull'. He gets worried, and so a policeman brought this to me and said, 'Is this the skull of the guy?' I looked at it and I said, 'No, it's not.' 'Well,' he asked, 'how do you know?' I said, 'You see that house dust there? That's where the skull has been sitting so long. That's not somebody that's modern. Then you feel how smooth that is on top? That's because people have played with it and so forth.' Policeman went back and found out that the guy that sold the old truck had been military in the Second World War, and had brought this skull back. We never found the skull of the body in the lake – it's out there on the bottom somewhere.

See, all of these have stories like that.

Tell me a bit about your own story – where you grew up, who your parents were and all that.

I was born in Staunton, Virginia. My dad was a lawyer, and he died when I was four, so I really don't remember him. He committed suicide in 1932 during the Depression. It was one of those things that were never talked about, but I did gather from my mother that he had invested money for some people, and when the stock market fell he didn't see any way out of it, I reckon. I don't know. The older I get, the more I would like to know. But you know . . . It happened.

My mother moved back to her farm, also in Virginia, and we lived there, I reckon, for four years, and then my mother remarried. She married my father's brother, so the name never changed, and most people thought that my stepfather was my real father. He was a geologist who ran two limestone quarries in northern Virginia, and I grew up in a little town called Stephens City.

When did you get interested in science?

Probably my junior year at college. I was a psychology major. The junior year I had the opportunity to take some electives; I took anthropology, and I see now that I was hooked from the first class. There was only one anthropologist in the University of Virginia when I was there, Clifford Evans, and I took every course he offered. I graduated in psychology, and later I went to college in Kentucky to work on a Master's in counselling. I still didn't realise that anthropology was what I wanted to do. The first semester I was majoring in psychology and minoring in anthropology, and the professor I had was a man named Charlie Snow. Charlie Snow taught me bones, osteology. I was at a stage in my life where I wanted certainties: if I learned that that's the femur, it's the femur everywhere in the world. Psychology was, 'Well, it's like this here, but it changes over there.' It was just concepts that kept moving. I needed something stable to latch on to, and when Charlie Snow came in one day and said he had an ID case and would I like to go, I thought, 'Great, I'd like to see how this is applied.'

I switched from psychology to anthropology and it was the best move I ever made. I got a Master's and went specifically to the University of Pennsylvania to work with Dr Wilton Krogman, who was known internationally as a forensic anthropologist. It wasn't called that at the time – they were called 'bone detectives'. The name 'forensic anthropology' came in the early 1970s, after I had gotten my degree. There were a number of us who were working with police and it was in the forensic area – 'forensic' means where science and law overlap – and we said, 'Well, hey, you've got forensic engineering, and forensic pathology, why don't we have forensic anthropology?' I think it was one of those ideas that matured in everybody's minds at the same time.

Okay, so when you'd got your Master's, what were you aiming to do?

I wrote my dissertation on skeletal material from the Plains area of the United States. I was working in the summers for the Smithsonian Institution, excavating Indian burial grounds, and so forth, and wrote my doctoral dissertation on prehistoric skeletal material. There wasn't a big jump between that and modern stuff.

When I came here to Tennessee, I established for the medical examiner a forensic response team, because I saw there was a need. Tennessee is mainly a rural state – there are about 85 rural counties, and a rural county does not have anywhere to put a dead body. A funeral home doesn't want the kind of bodies I get. They would put them out in the storage area because they smell so bad, and nobody wants to handle them. They didn't want them in police cars because

97

of the smell, and so I thought, 'Well, we need to establish a forensic response team.' Now we have a list of students who are on call at all times, and they go out and fetch bodies from anywhere and bring them here. And they work with the police to do the analysis.

Okay, back to your Plains Indian work – what were you doing with the Smithsonian?

Well, I had always liked Clifford Evans, the professor I had at Virginia, and I wrote to him when I went back to school in Kentucky, saying, 'Hey, I was one of your students at Virginia. I have found that my interest is in physical anthropology, skeletal remains.' I got a note back from him immediately, and he said, 'The archaeologists have been excavating lots of sites, and they've been finding human bones in many of them. These human bones have built up in the physical anthropology section of the Smithsonian, and we are looking for somebody to study them.' My God, this is just what I wanted to do, and here was the opening! So starting in 1956, I worked for the Smithsonian every summer until 1960.

And what were you looking for?

In the first two summers I analysed the material that had been dug up – just boxes upon boxes of bones – and I wrote reports and so forth. Then they were digging a couple of large Indian villages in South Dakota and the archaeologists said, 'We need somebody to dig the burial sites – why don't you send Bass out here?' So in the summer of 1958 I went to the field. The Army Corps of Engineers were damming up the Missouri River for a reservoir and they were trying to recover as much of the prehistory as they could before it was flooded. I amassed a collection of about 5,000 skeletons over those years.

I spent 14 summers out there, digging in the Plains. Best summers I ever had. Absolutely hard work – hot, dirty, all day fighting the elements. You get violent storms, tornadoes and things like that, but I never had anybody hurt.

How did you move on from the Smithsonian?

There were two jobs available that year in the physical anthropology area: one in Kansas and one at the University of Utah. I went down to Kansas, and they liked me and I liked them, so I started there in September of 1960. But I continued the research in the Plains – I was out almost every summer in the '60s.

So when did you start to do police work as well?

Well, I started with Charlie Snow in 1954. And I did a number of cases with Krogman in Pennsylvania. Krogman had an interesting teaching style: he would take me out on a case and he wouldn't say anything about it at all. He would look at it; I would look at it, and then he'd say, 'Okay, Bass, is it male or female?' You could not look it up, and he'd say, 'Come on, guy! How old is it? What's the race? What's the manner of death?' And we would discuss the case. Damn, that's a good way of learning! And when I started teaching, I did the same thing. I always took my students out. 'Okay, I want you to look at that now. While we're taking the pictures of this for the police, you determine: what's the age, what's the sex, what's the race?'

So a lot of it really is observation and learning from experience?

Oh, absolutely. You have to be an observer of minutiae – to look at the little things. And you have to know what those little things mean, so that's where your academic training comes in.

We had an interesting case that I did with some students over in Cumberland County about 60 miles west of here. This was a 16-year-old girl from Knoxville who was missing for two years. We get over to Cumberland County . . . It's difficult to look at bones and tell whether they've been dead a year or two years. We're looking at the skeleton, and lo and behold, before we even pick the skull up we're looking through the foramen magnum – that's the hole in the bottom of the skull through which the spinal cord enters – and we can see a wasp nest. Okay, so what does this tell you? To be a forensic anthropologist you've got to know a little about a lot of things. When do wasps build their nests? I had to look up the entomological literature: wasps build their nests in Tennessee in May and June. We are now in March and we find this wasp nest in the skull, so that means you've got to go back to the May or June before. Now wasps aren't going to build their nests in there unless it's dry, which means you've got to go back to the year before to give it time to decay. This is how we knew – while we were at the scene – that, hey, this is the girl that's been missing two years.

So, Bill, what is your motivation in doing what you do?

A challenge to see if I have the knowledge to figure out who this individual is and what happened to them. And the satisfaction of knowing that I've helped a family, or helped society put away somebody that probably should be put away.

Those are the things that drive you?

Yes. And students. I have two families: I have three biological sons, and then I have my master's and doctoral students, and we're very close. I have for years helped students pay their bills in college. There are a lot of my graduate students who will tell you, 'I wouldn't be here today if Dr Bass hadn't given me a place to stay at home . . .' You know, when you've got a good student, you don't want them to flounder because economically they couldn't get it together . . .

Tell me, how does the medical examiner decide whether to call in a forensic anthropologist or a forensic pathologist?

Well, it depends on their personality. If you're an older pathologist and, as I always say, you got your degree from God, then probably *I* would never be called in! If it's a younger pathologist who realises, 'Hey, Bill Bass knows something I don't know,' then I'll be called in.

What new tools have come along during your time that have really made a difference?

The one you think of right away is DNA. I mean, we didn't have DNA 15 years ago. And then there have been increases in all of the technologies. My book *Beyond the Body Farm* is about cold cases – cases that have taken 20 to 30 years to solve – and the question behind the stories is: how has technology in forensics changed? There are chapters on a whole bunch of things, from sonar and scanning electron microscopy to DNA. We brought in the techniques that are used in forensics today to look at how they had changed from when I first looked at the cases. One, for example, talks about fluorescent light. Theoretically, your body chemistry is different from mine. If you take fluorescent light – this is what geologists used to study minerals – and you shine it on bones from different individuals, the bones will fluoresce different colours. I did that on a case from 1974.

A young lady, Liz Wilson, was killed in the fall of 1974, found in the early spring of 1975, animal scatter all over, in a field that had been mowed, and they'd only found 14 bones. They did find the skull, and she was positively identified from the teeth. The question was: were all of these bones from the same individual? How can you tell? Dogs had chewed them, and they didn't fit together, so I thought: well, let's use this technique of fluorescing. Sure enough, they all showed the same colour. I thought: well, we've got to check this. I had brought a few skeletons with me, and I used a two-year-old child's and it fluoresced differently from Liz Wilson's. We had a teenager from Tennessee and, sure enough, they were all three different. And so on that basis I said, 'Yes, the bones are all from the same individual.'

Turning to philosophical matters, you have to confront death almost every day of your life; what are your personal feelings about it?

Well, I don't know how to answer that. I don't like death, as I told you. When I die I don't want anyone coming to mourn for me . . . I think I will be cremated, and I will donate my cremation to the collection here – I don't think I'll stop being a teacher just because I die.

I've been kind of impressed with cremations and have done research. Most of my stuff for law enforcement agents now is the identification of cremation remains. There's a lot you can tell from cremation, if you know what to look for. I've made positive identification of individuals from the dentition that's been in there.

What else can you tell from cremation remains?

Well, you can tell whether it's male or female; you can tell the age of the individual from burnt bone fragments. And you may be able to tell race, but race is a little more difficult.

The burnt bone is pulverised. The pulveriser is about a five-gallon bucket with an axle in the bottom and a blade that cuts both ways, and the manufacturer guarantees ashes in 60 seconds. But if you go through those ashes you'll find there are still things that you can identify. You can identify parts of the skull because you have the suture line. You will get the articular surface of a bone, and they are all different, so you can get an idea of what bone it was. These are the things you look for – the minutiae that the average person doesn't see.

In the United States today people are harvesting parts of bodies – hips and legs – and sending them off to biological houses that grind the bone up; it's used in orthopaedic areas where you need some bone to fill in. Most of the individuals from whom body parts have been 'harvested' – and this is a racket; there are big lawsuits in the United States right now – are then cremated, so it's difficult for families to know whether they got back all that they should have got back of a body. I've looked at four or five cases where families wondered whether they'd gotten back everything they should.

But how can you tell that from what remains?

Well, sometimes – like one guy who didn't have any teeth at all and there were teeth coming in with the ashes – it doesn't take a rocket scientist to figure out: this ain't the guy! Now, it's more difficult if you have the cremations come back and they're in the low end of the weight range from what you'd expect.

There are very few articles on cremation weights, so one of the faculty members here and I did a study on that, and we have the largest sample ever reported. What I try to do is find out how tall the person was, because cremation weights are not based upon what you weighed when you were alive – the soft tissue burns off. Cremation weight is based upon whether you're big boned or small boned. The problem, though, is that, if you were big and fat, your bones are denser because they had to hold up more weight. And if you had osteoporosis, then we have another problem . . . But that's getting into a grey area. That's for another student to study someday.

Apropos of research, I read somewhere that your body farm doesn't just give you answers; it poses lots of questions too. What are the questions it's asking at the moment?

Okay, a question that intrigues me is: people with cancer who've had major chemotherapy, do they decay at the same rate as non-chemotherapy patients? We don't know; nobody's done that study. I have noticed, even though I don't have a very good sense of smell, that they smell different. You would expect that, because they've had all these chemicals put into them. But do they decay at the same speed? Do the maggots pick up these chemicals? We would think that they do. We're right at the stage where we can tell you what drugs people were on – marijuana, cocaine and things like that. If you want to find out what that individual was taking, you collect the maggots, grind them up into a 'maggot cocktail' and analyse them and you can tell what those maggots were eating. But we have not done this with chemotherapy yet – that's the real cutting edge.

Fascinating! You said that you can tell the race from the bones – how?

Okay, race is best determined in the skull. Caucasians, or whites, have narrow noses; blacks have broad noses. Blacks have a protrusion of the bone underneath the lips known as prognathism. Caucasians have a sill, a little 'dam' at the base of our noses on the skull; in black individuals you don't have that. It's just a straight shaft from the bone just above the teeth right into the nose.

One of my doctoral students is a woman named Emily Craig, who worked for 20 years in a sports medicine clinic in which they were dealing with knees and so forth, and she said that the orthopaedic surgeons would much rather operate on the knees of a black individual than a white. And the reason is that there's more space between the knee cap and the joint. Emily can tell in 90% of cases whether it's a black or a white individual just by measuring that angle taken from an X-ray of the femur.

And is race a controversial issue?

The controversial issue is that some people don't believe there is such a thing as race – mainly the cultural anthropologists! But race is a biological concept, not a cultural concept. Enough said. There are differences between people. I remember 30 or 40 years ago, those of us who believed this were shouted down by those who said, 'Oh, we're all people . . . Everybody is the same.' But differences between people do exist; we're not alike.

I mean, you have the Black Plague: many people died but some didn't. Why? Because they had some special genetic characteristic that allowed them to survive. The two wives that I had, neither one of them had the genetic ability to overcome cancer. My second wife, really a sad situation . . . Her first husband was a chain smoker, and she lived for 30 years with second-hand smoke. We were only married three years before she died of lung cancer, and she'd never smoked a day in her life.

Bill, one of the things that really strikes me is that all of the bones in your collection, they're a story to you, they're not just dead remains.

That's true. Particularly the forensic ones – you work on them, you get to know a bit about the people, and you think about: where's the rest of that skeleton? As I said, the forensic cases are a puzzle, and one of the things you want to do is complete that puzzle. You want to have all the pieces together so that you can say, 'Now I see the whole picture.' I think that as a forensic anthropologist you're never really satisfied until you've finished the puzzle.

CANCER IN CLOSE-UP

Christopher Fletcher
*Professor of Pathology, Harvard Medical School; Director
of Surgical Pathology, Brigham and Women's Hospital; and
Chief of Onco-Pathology, Dana-Farber Cancer Institute,
Boston*

Chris Fletcher grew up in Yorkshire. His father was a surgeon and his mother a social worker, and he wanted to be a doctor from as early as he can remember. 'But my father was not impressed when I went into pathology,' he says. 'Being an old-fashioned surgeon, he regarded the pathologist as one of the surgeon's handmaidens.' However, pathology was very well taught at St Thomas' medical school in the late 1970s, and Fletcher came under the influence of some exceptional teachers. They included Michael Hutt, who had set up one of the first medical schools in Central Africa, in Uganda, and who continued to provide specialist pathology services for this and other hospitals in Africa by post. 'We used to see astounding pathology that you would never ever have seen in Central London. It was very inspiring,' says Fletcher.

Chance led him to specialise in soft tissue tumours, or 'sarcomas', which are among the most difficult cancers to diagnose and to predict the outcome for the patient. In 1989 he set up the Soft Tissue Tumour Unit at St Thomas' Hospital, a pioneering consultancy service in London, which became embroiled in NHS politics as the government introduced the idea of an 'internal market' where specialists had to charge each other for their services. The new deal caused much antagonism, and in 1995 he left to take a chair of pathology at Harvard University in the US. Renowned for his skills in diagnosing sarcomas, Fletcher continues to get 'an endless stream of weird lumps' referred to him from all over the world.

———•·•———

What attracted me to pathology, I think, was realising that it is the underpinning of so much medical care. I mean, you're the one who's making the diagnosis. Very often you're the one who's actually telling the surgeon, or the medical oncologist, what they need to do next. They never tell the patients that, but half the time we're 'guiding' them in their clinical decision-making. Of course they're the infallible alpha males sitting there with the patients, but actually they've often got no fuckin' idea until we tell them what it is!

There's a wonderful statistic – and the problem is I don't know where it comes from – that something like 70–80% of clinical decisions in a hospital are based on either laboratory or radiology results, and those account for only 7–8% of the hospital budget in any given city. In other words, value for money is huge in terms of impact on clinical care.

When you began your career as a pathologist in London, what kind of 'patient profile' did you have?

When I started out as a trainee, it was just the same as any large hospital, and I then developed an interest in soft tissue tumours – things in limbs and retroperitoneum [the space behind the abdominal cavity] and chest wall, etc. I continued to do general stuff, but that was my research interest. Then I got a grant from the Cancer Research Campaign [now Cancer Research UK] to establish the Soft Tissue Tumour Unit, and we set up a clinic at St Thomas', with an orthopaedic surgeon, me and a general surgeon. I used to go and see the patients with them as well, and patients started coming from all over.

Can you remember what first interested you about these things?

Well, when I was a baby trainee at St Thomas' someone came from the skin world to the dermatopathologist there and said, 'Will you please write a review article about how you diagnose sarcomas?' (Those are the malignant things in soft tissue.) He went looking for a trainee, and said, 'Okay, somebody's going to write this . . . You can be first author; it'll be good for your CV.' And for no special reason I said I'd do it. I got very interested, and what was going to be one review article became a series of four, which were well received. And during the course of that I read and looked at lots of cases that I found in the files. I discovered that I liked it; we started doing little projects, and it just went exponentially from there.

So what was the state of knowledge of soft tissue tumours when you first became interested in them?

Well, it hadn't been a real focus of interest. There were one or two internationally recognised experts. But there was no molecular biology at that time, no genetics. And because soft tissue tumours are relatively uncommon, compared with things like breast cancer and lung cancer, people had difficulty gathering large enough numbers of cases to be able to work out properly how they behaved, or who was affected by them. Through getting consultation cases, you gradually build that stuff up. I got my first consultation case in 1987, and by the time I left England I was getting about 1,500 cases a year, which in the UK was huge. And I now get 4,500 a year. So you get enormous numbers of rare things, which enables you to put together definitive information on how a certain tumour behaves.

What exactly is 'soft tissue'? Surely breast and lung are also soft tissue, so why are they not included in this category of tumours?

Well, these are tumours of what we call mesenchymal cells. So it's cells that make the connective tissue – the supporting tissue like smooth muscle, skeletal muscle, adipose [fat] tissue, nerves, blood vessels, all those things. That's what the pathology world means by 'soft tissue'. And of course all of those things are ubiquitous, so you get tumours that are composed of mesenchymal cells in almost any anatomic site. You can get a blood vessel tumour just as well in your heart or your leg or your bowel as anywhere else, because everywhere has got blood vessels.

And how common are these things?

Well, world over, there's about the same incidence for malignant ones as for testicular cancer and Hodgkin's lymphoma, which is about 15 cases per 100,000 people a year. But for every malignant one there are about 100 benign ones. There are lots and lots of benign things, and when one looks down a microscope one doesn't necessarily know intuitively whether something is benign or malignant. Soft tissue things are notorious for the difficulty of separating the two, and our referral practice became the biggest in any sort of pathology in Britain at the time. For some reason sending cases as referrals or consultations hadn't been such a big thing before. Maybe people never used to worry about rare things. Or they would show them to their pals down the hallway, who'd say, 'Well, it's probably such and such,' and that was good enough. I don't know. But in the 1980s people began seeking second opinions for difficult or weird things, and soft tissue is over-represented in that regard. And that's when things became problematic for me.

This was the time when Margaret Thatcher was trying to make healthcare market-driven. The idea was that if a patient went from Brighton to London to have a hip replaced, the health authority in Brighton had to pay the health authority in London for doing it. I ended up with this large consultation practice – 1,500 or so cases a year – and some administrator got the idea: 'Here's the perfect example of market-driven. There are pathologists around the country saying, "I need to send my cases there." So what we'll do is ask him to send bills for his services, and if what he does is worth while the health authorities will pay for it. And with that money he'll be able to pay his support staff.'

Remember, nobody had ever sent a bill for a referral before. And it was frightful – people went berserk. The government folks provided me with all these forms that I was supposed to make people fill out. And then, because we were part of the European Community, they said, 'Oh we can do it Europe-wide'!

I got crucified by my colleagues, who said, 'How come you're charging for consultations when nobody else does? Fuck you, I'm never sending you a case again!' That sort of thing. It became quite a little *cause célèbre*, and it ended up being on television. Channel 4 made a documentary for the *Dispatches* series about a group of specialist services that were being crucified, and I was the pathology one that was picked. There was a bone marrow transplant service for children at the Westminster, and they had an oncologist who said, 'People can't get round the bureaucracy to enable them to send their child to us, so children are dying . . .' Of course, it was dramatised for the sake of television. It was in the newspapers too.

So what would have happened if you hadn't charged?

They wouldn't let me do the work. It was very simple. They said, 'You're not going to spend your time doing that unless people pay for it.' And, of course, nobody did; we collected peanuts.

So did you find the number of cases being sent to you dried up?

Well, it was a funny combination. I was getting better known, so it was on an upward trajectory. But some hospitals said, 'We're not going to engage in this,' so it would probably have gone up faster if it hadn't happened. And the hospital put the people who worked for me on monthly renewable contracts because of the uncertainty of my financial situation.

How awful!

Well, it *was* awful. After about two years the administrative people began to realise the system wasn't working, and in the end they turned a blind eye and just assumed I was going to lose money. But they never took away the implicit threat.

So were you able to go on practising good stuff during that time?

Well, yes, because I got thicker and thicker skinned. I started writing in some of my letters, 'I'm enclosing this form as I am required to do. Please return it if you can . . .' I phrased it in a way that said, 'I know this is bullshit,' without saying that explicitly. It started out being terribly stressful, because one thought: 'We're just going to be closed down.' Then one realised that most people thought it important, and that this was more political than anything else, and in the end I just gritted my teeth and thought: 'Fuck it!' But when Harvard tried to recruit me, it wasn't difficult to decide where I'd be better off.

Absolutely. But tell me about the development of soft tissue tumours as a speciality, and your growing expertise in this field.

What you're really saying is: what's changed in these 20 years? I see so many it's hard to think back. Nowadays we average 25 new cases a day from all over the world. And remember, these are cases where pathologists have got stuck – in other words, they're disproportionately difficult. So it's somewhat divorced from reality because you get such a biased subset. But what we've achieved over the 20 years is we've defined or described for the first time lots of tumour types that were not formerly recognised.

Like what?

Well, as a general principle, it's things like tumours that were formerly thought to be malignant and you get enough cases and follow them up and discover they're not. Things that were formerly confused – you know, benign and malignant lumped together – we've split them apart.

Probably the single biggest achievement concerns something called malignant fibrous histiocytoma, or MFH. When I was a baby trainee in the 1980s, this was the commonest cancer in the soft tissues, and had been for the preceding 20 years. It was supposed to account for 65–70% of adult sarcomas, but we showed over a period of time that MFH actually doesn't exist! It's like lumping different kinds of dogs together, calling them all 'Mutt', and losing the fact that some are Pekingese and some are Irish wolfhounds, etc. These tumours all behave differently and respond differently to treatment. But that has been a very hard thing to overturn, because, just think: patients all over the world have been diagnosed with something that generations of pathologists firmly believed in.

It bothered me that no two cases looked quite the same. They shared some characteristics but, in sub-classifying them, you found ones that were terribly aggressive versus ones that weren't particularly; ones that were very chemosensitive versus ones that were not, and they just had very different outcomes. We published a paper in 1992, and in the 2002 World Health Organization classification it was acknowledged that MFH doesn't exist. Now everybody is getting to learn this, and it has had an impact on patient care, so it's worth doing.

So before that, how did they used to treat these tumours?

Well, surgery plus some radiation or chemotherapy. But doctors wouldn't be able to sit with a patient and say, 'You're in trouble' versus 'You should do okay.' I mean, they were all just lumped into a single diagnosis which meant: '50 per cent of you die and 50 per cent of you don't, and we don't quite know which . . .' It was just sort of dogma.

And how were you able to tease them apart?

A lot of that was done by looking down a microscope at lots of cases, and then using this technique, immunohistochemistry, and electron microscopy. So, fairly ordinary tools.

So it was literally just the number of times you saw these things . . . ?

Observational mainly, yeah. And then employing tests to try and objectify that, so it's not witchcraft. It's, 'I think these cells may be showing a certain line of differentiation, so let's see if they express these proteins to support that,' or whatever. And then, 'Let's see if that means anything in terms of behaviour.' It's clinical research, it's not basic research, so there's no question it has direct patient impact.

And it's clinical rather than basic research that you've been involved with most?

Yes. And loved it. Still do!

You say you get referrals from all over the world – are you seeing different things from different geographic locations?

Not dramatically different, no, though there are some differences. This is not my area of expertise, but, for example, gastric cancer and colon cancer are significantly different in Japan from here. Those are driven by diet and environmental things. With the soft tissue tumours that I deal with there are no data to suggest they're environmentally influenced.

What do the data show?

That we have no clue what causes most of them! Just no clue at all. There's a subset that are associated with radiation – usually for a prior malignancy, not because you live next door to a telephone pole or something. That probably accounts for 10% of them, and that's interesting, because we're seeing more and more of these. Nowadays women tend to have lumpectomies and radiation for breast cancer instead of having mastectomies, and we're seeing more and more radiation-induced vascular tumours of the skin of the breast. Try to tell the radiation people this and they don't want to hear – because again there's so much dogma. And breast cancer is such a politically sensitive issue. But it's quite striking: as I travel around the place giving talks and meeting people, it's quite clear that everybody's seeing the same thing.

And do these tumours then go on to spread?

Some of them are malignant, yes, some of them are scary. But that's a good example of something we'll probably have to follow for 10 years or so to see how many of them actually progress to being frankly bad.

The problem is that it's a 'swings and roundabouts' thing, and it's something nobody ever talks about in medicine. You know how everyone proudly announces that they cure children of most of their cancers nowadays? A huge proportion of them come back in their twenties and thirties with secondary malignancies induced by the treatment they had when they were little. But that bit doesn't get talked about so much. It becomes a moral or ethical thing because, after all, if you've given someone 20 years of life, most people would say, 'Well, that's not a bad outcome, is it?'

What sort of secondary cancers do they tend to get?

By far the commonest is what's called myelodysplastic syndrome, because you've whacked the bone marrow with chemotherapy – the screwed-up bone marrow predisposes to all sorts of leukaemias. Of all the little children that get radiated, a significant subset will come with a radiation-induced cancer years later.

Nowadays, too, the thoracic oncologists are seeing radiation-induced lung cancers in patients who had breast cancer treated 25 years ago and survived. Again you might say, 'Well, hey, if they were going to die of breast cancer 25 years ago, it's not a bad deal, is it?' But it's not something that gets talked about much.

I've heard others say that it's very difficult to challenge the status quo – sometimes they can't get papers published if they go against received wisdom. Have you found it difficult to get your voice heard when you've come up with challenging information?

Well, the thing I talked about earlier, MFH, it was very hard to get that published. When you submit things to most pathology journals, the norm is for there to be two peer reviews, then an editor decides whether or not to accept it. It took six reviews, I think, before they accepted the MFH paper, and then it was 18 months before it was published. I think things have changed a bit, though. Diagnostic surgical pathology has become more objective during my career time; it's more reproducible, evidence-based. And I consider this a strength, because one of the things that led the basic science guys to be critical of the clinical people in the 1960s and '70s was that a lot of the clinical stuff – before the introduction of technologies like electron microscopy or molecular genetics – was a matter of opinion. It was almost who had the biggest balls – you know, 'I say it's so and so,' and if you're big and famous, then it *is* so and so!

There was an awful lot of dogma. You'd go to meetings where famous people would sit and argue. But there was no right or wrong, and it made me understand why basic scientists would say, 'This is all bullshit.' Now, it's much, much more objective.

But it's interesting. You'll meet people who'll tell you that molecular genetics are going to objectify things even further, and then, 'You won't need people like Fletcher any more, because we'll be able to decide by a pattern of gene expression, or a chromosome translocation, that it's so and so.' But what's fascinating is that the more genetic information that's discovered, the more you find that completely different tumours have identical genetic signatures.

Really?

Oh yes. There are wonderful examples of that – which the molecular genetics guys don't like to talk about. That's another thing that gets pushed under the carpet.

For example, ETV6-NTRK3 is a gene fusion from a chromosome translocation that was thought to be specific for a thing called infantile fibrosarcoma in children. It's a tumour that grows in the limbs that looks hideous, and you give it chemo and it disappears. Exactly the same chromosome translocation has been found in a specific type of breast cancer that is actually rather aggressive. And in a type of leukaemia. So in other words, the same initiating chromosomal event, depending presumably on what cell type it happens in, gives you three completely unrelated tumours.

Goodness! Did you discover those things?

No, but we have discovered some similar things in the more recent past, but that's the most dramatic example I can think of. What we're also seeing are things that look benign and banal down the microscope that then metastasise and kill people. And you say, 'How is that possible?' In fact, most pathologists, when you tell them that, say, 'Well, they're obviously misdiagnoses. Somebody screwed up.' But one realises over time (and it comes back to genetics again) that although these tumours look the same down the microscope as quite common things, it's probable that they have had some other genetic event that says, 'Okay, I'm capable of getting into a blood vessel and spreading now,' which we don't know how to recognise. And only if one gathers outlying cases could one ever think, 'Okay, let's find out what's different

genetically or molecularly between those and the other ones.'
So even if the aggressive type is *ridiculously* rare, they may
provide valuable information for understanding the whole
process.

*So it's a cumulative thing. You need a great library of
knowledge and samples, I presume?*

Oh, absolutely. You'll have seen people with little brown
lumps on their lower legs? They're called dermatofibromas,
and they're fantastically common. They're banal and benign,
and in fact some people believe they're not even tumours. It
was realised about six years ago that once in a blue moon
these things can actually spread and kill you. And with the
first few reports you could sense disbelief. People would
write letters to the journal saying, 'This case is clearly a
misdiagnosis. I can't believe you published that,' and so on.
We now have 12 or 14 of the ones that have gone wild, and
we'll have to try and work out how they differ from the
benign ones. But only by gathering a reasonable number,
with follow-up, and then working with them will one ever
discover what's happened.

*So do you find, the longer you are in pathology, that the
more complex it gets, and the more you realise you're seeing
differences the whole time?*

Yes. And my goal is to try and get other pathologists to see
that – and also to be brave enough to educate clinicians. There
are personality types that go with specialties, aren't there?
Surgeons tend to be gung ho bullies (major generalisation!);
orthopods are big, grinning people with large hands who
aren't that aggressive; psychiatrists are kind of weird; and
paediatricians are highly strung. I think pathologists are
some kind of combination of shy and retiring, or mentally

attuned to being in a service role. In other words, we wouldn't have a job unless people sent us tissue, and we're doing our work for someone else – we're not the ones who actually tell the patient it's benign or malignant (although when you end up being a specialist, you have more and more patient interactions, but that's an oddity). I think a lot of pathologists don't see it as their role to educate clinicians about the uncertainties or unpredictabilities, whereas I must say I feel the reverse. I feel quite strongly about it, and I'm sometimes a bit outspoken! But it works. In this hospital, pathology is a highly respected specialty.

Returning to the development of your career – you ran your clinic in London and got strangled by the bureaucracy, and then what happened?

This department at Brigham and Women's was looking for a director of surgical pathology, I got asked to interview and was offered the position. And that's another wonderful thing in medicine – if you're good at something, then you get hired to be an administrator, and you spend all your time in committees! So I came here, and now my main job is administrative. We have a huge department: 75,000 surgical specimens a year; 60-odd faculty who work for me; and 40-odd trainees. There are more than 100 doctors who are all related in some way to surgical pathology.

Do you mind doing administration rather than looking down a microscope?

I don't mind. It's just that there aren't enough hours in the day to do everything. Now, as you 'grow' young ones around you, they can become the front end of some of the academic productivity. I can give projects to junior faculty or residents for the research things, and we can do them together, because I don't get time to write as much as I used to.

Do you have some good juniors?

Oh, wonderful. This is regarded as one of the strongest departments in the country, so there are some fabulous people. Seventy per cent of our trainees have PhDs in addition to their medical degrees before they even start in pathology, so they're very impressive.

What are you researching at the moment?

God, we have about 20 or 30 projects going at any time! They range from describing more tumour types that we've identified – things that people had never imagined, which are going to cause a certain amount of controversy.

On a quite separate note, here's a really banal example: most people have traditionally thought, naively, that tumours arise from their normal cellular counterparts. So if you've got a smooth muscle cell, then you can have a smooth muscle tumour; if you've got a fat cell, then you can have a tumour made of fat cells. But in the soft tissue things I've been talking about, the truth is we've no idea at all what cells they arise from. Most tumours showing skeletal muscle characteristics arise at sites where there isn't any skeletal muscle. So the idea was obviously nonsense all along.

My humble belief is that you can take almost any cell type and if it undergoes a kind of genetic catastrophe that says, 'You can go bad now,' it can then reprogramme and become something different. Because, remember, every one of your cells has got your entire genome in it. Your fingertip has got the gene in it that made your pituitary gland as well. It's just that during the development of the embryo that gene was switched off because that was going to make a finger cell, not a pituitary gland cell. But the information is there.

I'll give you an example – nothing as bizarre as growing a pituitary in your fingertip! Anywhere that you make secretions, like the breast or salivary gland, you have ducts, and around those ducts are things called myoepithelial cells. They're the things that actually make the ducts squeeze so that saliva or breast milk or whatever comes. Normal myoepithelial cells are only ever found around ducts in organs like that, but we started finding myoepithelial tumours in all sorts of ridiculous places – in deep soft tissue of people's legs and things – where there are no ducts of any kind. Such tumours must always have been around, but it's just that people didn't recognise them because, you know, 'You can't possibly have a myoepithelial tumour there.'

We started describing those in the late 1990s, and it was like a revelation to people. I was then sent, within three or four years, 100 more cases of something that didn't exist before in theory. Of course it existed; it's just that none of us were open-eyed to it. We now have way more than 100 cases, and we've gathered a whole series of malignant ones, in little children, that are very aggressive.

So what happened to children with these tumours before they were recognised?

Well, the children died, and they were probably just forced into categories that weren't correct because the pathologists couldn't think of anything else.

This identifying of new things keeps happening all the time. You see something and you think, 'That looks weird; it reminds me of these other things . . .' And you put them away in the back of your mind. Over a period of years you think, 'I'm sure we've seen seven or eight of those now,' and then, thank God for computer databases, you can find them relatively easily. In the old days you had to look through all those handwritten little books. [*gestures towards a row of hardback exercise books on his shelf*] Usually when I start recognising an entity, I'll call it something like 'distinctive gelatinous thing'. Sounds ridiculous! It's just a way of remembering something when I don't know what it is. So I'll get out all those things that I called 'distinctive gelatinous thing', and it's a wonderful sensation, because nine times out of ten they all look exactly the same. Then you say, 'Okay, let's find the follow-up and work out what they are.' Then you present the stuff.

It's very pleasurable, a real thrill. And when it's clinically useful – I mean, to me, that's what matters.

For example, we've got a thing we see in young patients that we haven't published yet. They develop multiple nodules, usually in one limb, in the skin, or muscle or bone. Lumps and bumps. If you use some fancy scan you usually find they've got 10 or 15 lumps. Down the microscope they look like a thing called epithelioid sarcoma, which is a nasty type of cancer, so we used to think they were some funny variant of that.

After a bit I realised I could recognise these things just down the microscope, without knowing that there were multiple bumps. Then I knew that was valid because I'd say in my letters, 'I've seen cases like this where they'd subsequently develop multiple lumps. Please scan the limb.' They'd scan the limb and say, 'Wow, you're right. We've got eight or ten lumps!' So that's the first thing that tells you you're on to something. You assume that this must represent a pattern of spread and therefore you intuitively think it must be bad. But, 'touch wood', we now have about 25 of these cases and not a single patient has developed distant disease.

So it just stays within the limb?

Yeah. So it's clinically distinctive, intuitively horrible, and yet turns out to be – I won't say 'benign' because I think we have to follow it for a long time. And we're trying to find out what the hell they are, because, I mean, they don't 'exist' right now. You could pull every medical textbook off the shelves and you'd not find it. It doesn't even have a name. We've had a couple of patients here with this disease, and I go and sit with the family and say, 'Look, we've seen 20 cases and they seem not to spread, but I can't promise you they don't.' Right now we've only got a few years of follow-up, and we've all learned that cancer can be a long-term disease. People used to say that if you survived five years you were cured, and, of course, it's absolute nonsense – lots of things come back later.

Do you still get a thrill seeing all this unusual material?

Yeah, absolutely. The best bit of the day is that. I get a lot of pleasure from running this part of the department. I like seeing the trainees do well, and I don't mind doing admin. But the fun bit is seeing lumps and working out what they are. And feeling as though you're helping patients – I mean, I think I'm probably a doctor at heart, first and foremost a physician. I think of pathologists as physicians, and I don't like pathologists who aren't willing to be engaged in clinical care. I think that's cowardly.

One of my most memorable visits from a patient was a lady from Wisconsin who just wanted to say thank you. She was probably 60-ish, and she and her husband were bikers – borderline morbidly obese bikers. So these two *huge* people arrived in this leather gear with studs all over it [*we laugh*] – and they were charming. I normally think bikers look kind of scary, and they were delightful.

Actually, I'd told her she had cancer, whereas somebody else had told her she had a benign thing, and she told me, 'If you hadn't said that, I wouldn't have got the right treatment.' It turned out I was right, unfortunately, because it spread. But she came to say thank you. Human beings are complicated things!

You obviously feel at ease with talking to patients – d'you think that's because you grew up in a medical household?

Probably. My father was quite traditional, so he would cart me and my sister round the hospital sometimes when we were little. Actually, in real life it was probably my mother saying, 'Can you take the kids out of the house for a bit?' Then on Christmas Day, when I was young, they would have surgeons carve the turkeys on the wards, and families were expected to show up too. They'd usually make the surgeons dress up, and we'd have to traipse round behind them – you know, the adults drinking endless sweet sherries and becoming progressively inebriated. When you were a small child it was very unnerving, but as you became a young adult you just drank the sweet sherry as well.

What would you say that being a pathologist has taught you about life?

Probably the two most important things are the recognition that life is very much a lottery, and that death can happen at any time. One of the problems I have now is that the demand for all of the things we've talked about is essentially limitless, and I have to work harder and harder. The job just gets bigger, and at some point I'll have to reclaim my own life, because I don't want to drop dead having done nothing else.

But no, what I've learned from the job is that, 'There but for the grace of God go all of us, day after day,' and what's painfully true is that we're all completely dispensable. If I were to drop dead tomorrow, people would find others to send their lumps to. One of the most famous pathologists in the world was chairman of this department, the man who recruited me, a man called [Ramzi S.] Cotran, who used to edit the standard textbook used everywhere across the world. Cotran had melanoma [skin cancer] and he died some years ago. Until he died, anyone looking to make a senior appointment almost anywhere in the US would call him up and say, 'D'you think we should hire so and so?' Cotran was Pope-like and very, very widely respected. And within this hospital, of course, he made pathology very powerful. For the first three months after he died, it was, 'What would Cotran have done?' and everyone would do the same. But after six months it was, 'Well he's dead now . . .', you know? After a year he didn't feature in the conversation at all. And now the young ones will say, 'Who's Cotran?'

And of course, it's true for every one of us.

Seeing so much of death, are you at ease with the idea of dying?

Good question . . . I don't worry too much about dying. From personal experience, having taken care of dying people, having talked to dying people and having watched one's own family members die, I've never actually had the sense that most people are scared when they get to that point, which I find reassuring. I'm not personally too troubled by it, no.

THE DEADLY SECRETS OF SPANISH FLU

Jeffery Taubenberger
*Senior Investigator in the Laboratory of Infectious Diseases,
US National Institute of Allergy and Infectious Diseases*

A serious preoccupation of public health professionals everywhere is: when will the next flu pandemic hit us? – as it surely will. In order to predict and prepare for such an event, we need to know what makes 'ordinary' flu viruses periodically become so efficient that they can zip round the world in weeks, killing millions. These questions began to obsess Jeffery Taubenberger, then working at the Washington-based Armed Forces Institute of Pathology, when he discovered in its massive archive, dating back to the US Civil War, cases of people who had died in the 'Spanish flu' pandemic of 1918, the deadliest yet. Could he and his team recover the virus from these ancient blocks and slides? And could they learn anything useful from them? This was the ultimate challenge to a pathologist schooled in the wizardry of molecular biology. But, he says, 'This kind of "viral archaeology" is extraordinarily painful! And very slow.'

Now at the National Institutes of Health and head of a brand-new 'high containment' laboratory, Taubenberger continues to ask pressing questions about 1918 and other influenza viruses, such as the current bird and swine flus. But he has been asking difficult questions of science since he was a kid.

At four or five I already knew I wanted to be a scientist. What fascinated me was the idea of figuring out how things worked. I ended up, as a kid, reading lots and lots of biographies of scientists and inventors – people like Thomas Edison, Henry Ford, Pasteur and Koch. As I got older I became interested in chemistry – rockets and fireworks and things that go boom – and my friends and I made all sorts of things that I wouldn't let *my* children do! We had a lot of fun with that.

Then I became interested in biology. Animate things seem so much more complicated than inanimate things, and the systems that control biology are just fascinating to me. So I started working in a lab of the National Cancer Institute as a volunteer when I was in high school – that would have been aged 13 or so.

Doing what in the labs at that age?

I was doing molecular biology, actually! At a time when molecular biology, in a sense, was being invented, I did a project in which I extracted DNA from a variety of sources to answer the question: is the total amount of DNA per weight of tissue related to how complex an organism is? I grew bacteria, and I took DNA from a plant, and from an animal (we bought liver at the butcher's shop), and then I calculated the weight of the tissue and how much DNA was recovered. The hypothesis is not correct, actually: there's no relationship between total amount of DNA per dry weight of tissue and the sort of 'complexity' of the organism. But anyway, that was my tenth grade, high school science fair project!

And what was going on in that lab was interesting. I met a man named Dr William Drohan, a molecular virologist who was studying endogenous retroviruses – that is the same class of virus as HIV, but these viruses can actually integrate themselves into the genome of cells and become resident, become part of your DNA. All humans and all animals have copies of these ancient viruses as part of their own DNA. They have been incorporating themselves into animal genomes for hundreds of thousands of years, or forever, basically.

And are they doing anything?

Well, that's a difficult question. Most of them are inactive. They have mutations such that they cannot actually make most of their genes. But they do move around, and they do have the ability to disrupt other gene functions. There was a big theory in the 1970s that many cancers are caused by viral infections, and that some might be caused by endogenous viruses – not the typical viruses that you're exposed to, but ones that you actually have encoded in your DNA.

I was working in William Drohan's laboratory throughout high school and he mentored me. When it came time to go to university I decided I wanted to stay in the area, because it was such a fantastic opportunity being in his laboratory. This was the late 1970s, when a lot of the basic techniques of molecular biology were just being invented. So I went to George Mason University, in Fairfax, Virginia, very close to my house, and I worked in that laboratory half-time during the fall and spring semesters and full-time in the summers for the four years.

And what did you do at university?

I obtained a Bachelor of Science degree in biology in 1982, and Bill Drohan strongly counselled me throughout my time with him that I should go to medical school and do a combined MD/PhD degree. So I entered a combined programme at the Medical College of Virginia, in Richmond, with the idea basically that this would augment my research career. I really didn't envision myself as a 'physician' at that point in time. But what I discovered, shortly after arriving in medical school, was that medicine was really fascinating! [*laughs*] It just sort of drew me in intellectually.

The rewards of medicine and the rewards of basic science are usually quite different. In a sense they're complementary. In medicine you can get satisfaction in a much shorter time period, in that you do something, there's an outcome and you receive satisfaction. To take a very simple example: somebody comes with an appendicitis, has an appendectomy and is saved, when otherwise they would have died. That's enormously satisfying – hence the egos that surgeons develop. [*laughs*] But that kind of instant gratification is very rarely, if ever, seen in science. You know if you have that sense it's probably because it's *wrong*!

You chose to specialise in pathology and did your residency at the National Institutes of Health in Maryland – how did your career unfold after that?

A person who was a mentor to me during my residency is a pathologist by the name of Timothy O'Leary. He was an excellent teacher and a great man. He knew a *huge* amount about autopsy pathology, but also was a person who had very basic science interests – he had a career in which he had clinical pathology contact but was running a basic science lab. I was very impressed by him. He showed me that my dream was possible: that you could still keep a hand in pathology, a hand in clinical medicine, but also do basic science. I was inspired by the fact that, although he was doing a lot of basic science, he was still an excellent and caring doctor.

Timothy O'Leary subsequently accepted a position at the Armed Forces Institute of Pathology, the AFIP, in Washington, which is one of the oldest and most venerable pathology institutes in the world. It was founded in 1862 by executive order of Abraham Lincoln during the US Civil War to study diseases of the battlefield. Pathology as a medical discipline was kind of being invented at that time, the mid-nineteenth century, with advances in the understanding of basic pathophysiology. So the AFIP was founded as an army institute, but it transitioned in the early twentieth century to become a centre of pathology expertise that also served the civilian world. The idea was that the Institute would serve as a consultative base for difficult cases, interesting cases, in pathology. People would send them cases that would then be kept in their repository as a permanent and ever-growing collection. This repository – of fixed tissues, blocks, slides and so on – has been growing since the 1860s to the present, so that they have millions and millions of cases.

And beautifully written up and recorded, I imagine?

Yes. They have computer-searchable archives for key words and diagnostic information, and they've gone backward in time to about 1915. And the way the AFIP's tissues are stored – in warehouses with these enormous racks – is impressive to see. They have automated machines that retrieve trays of slides: you punch in the number of the slide you want, the machines move and out of tens of millions of slides this one tray comes shooting out.

Tim O'Leary took a position as chairman of the department of cellular pathology at the AFIP in, I think, the late 1980s. He wanted to create a molecular diagnostic pathology component, with the idea that molecular biology tools would be useful as an adjunct in anatomic surgical pathology. He was looking for people who had dual interests in anatomic pathology and molecular biology, and he recruited me to come and head this new group.

That happened in 1993. It was a natural progression for me, and it ended up being a fabulous niche. Our task was to develop tests that could be used on typically processed surgical pathology or autopsy pathology material. You want to extract genetic material – that is, DNA and RNA – from a tissue that has been fixed in formaldehyde and embedded in paraffin, but while this fixing process preserves the *structure* of the tissue, it actually does terrible things to the DNA and RNA inside: chews it up and causes cross-links and all sorts of bad things. So we had to develop techniques that would allow us to reliably get nucleic acids (DNA and RNA) from such fixed tissue, and then figure out ways we could ask the kind of diagnostic questions we wanted to ask. We became the only facility, at least initially, that could do these kinds of tests, so people would send us cases from all over. It was an interesting time.

So how did your interest in flu arise?

Well, we were doing, in my lab, all kinds of different projects that linked around the idea of starting as a pathologist and asking: what could you do using molecular biology tools? And at some point I read an article in *Science* magazine in which some very clever people had done an analysis of DNA extracted from John Dalton's eyes to prove that he had classic red/green colour blindness. Because it was John Dalton – the very famous chemist, the person who came up with the modern atomic theory of matter – the eyes were preserved in mason jars after he died in the 1840s. These researchers got permission to take a tiny piece of an eye, and since the mutations associated with colour blindness are now known, they analysed his DNA and showed that he actually had Daltonism [*laughs*], which is the old term for colour blindness.

At the AFIP we had a weekly Journal Club, where each of us would take a turn finding an article in the literature, presenting it and discussing it. I presented this at our group and it just made me think: I'm sitting on a collection of millions of cases going back over 100 years. Maybe we can find someone famous in our collection and do something kind of cute like that. I had a brainstorming session with Tim O'Leary, and we came up with the idea to go after the 1918 flu. It immediately resonated with me as a great idea: rather than just a trick like the John Dalton case, which was clever but didn't really advance medicine, the 1918 flu would be enormously useful, and potentially of practical importance.

The 1918 flu epidemic was something I'd heard about briefly in med school. It was a huge outbreak that killed millions of people at the end of World War I. Numbers keep going up as people look deeper into the records, and the current thinking is that 40–50 million people worldwide died in the space of one year of influenza, which is an absolutely shocking number – you'd have to go back to the Black Plague to find a disease outbreak that would kill so many people.

I had a molecular biologist, Ann Reid, working in my laboratory, who expressed an immediate interest in this project and wanted to work on it, so together we started reading up whatever we could. We read a book by the historian Alfred Crosby called *America's Forgotten Pandemic*, which described the outbreak predominantly in the United States. He was intrigued by the idea that here was something that happened only 80 years before but that basically no one had heard of, whereas every elementary school student has heard of the Black Plague. His basic question as a historian was: how could you have an outbreak in which tens of millions of people die, and yet it doesn't become part of the cultural memory? He doesn't have a definitive answer, but he speculates it's because it was too fast: it all happened within one winter season, it was the end of World War I, all this horrible stuff happened at the same time, and people just got on with their lives afterwards and didn't think about it.

Anyway, this started literally as a hobby project, an idea that was really interesting, but the chances of success seemed extremely remote.

Can you describe how that flu virus behaved? Because people are always saying, 'I've had a touch of the flu,' but that was something quite different, wasn't it?

Right. This virus seemingly came out of nowhere, and it caused disease that was recognised at the time as influenza. Now, influenza was not known to be caused by a virus at that point – though the idea that viruses existed was beginning to be accepted in the scientific and medical literature. Virus, of course, just means 'poison' in Latin. They knew a lot about bacteria by 1918, and they were able to culture and identify a large number of bacteria. They knew how big they were, and they developed filters that should block the passage of all the bacteria they knew about. But, starting with experiments in the 1890s, they found that you could filter out everything that should be infectious, and yet the liquid coming through at the end was still infectious. So they had the idea that what was infectious was a *chemical*, a poison, a 'virus'; it wasn't actually an organism. They had no electron microscopes at that time, and you couldn't see a virus with a light microscope, so they had no way of knowing really what a 'virus' was. There was just this infectious 'thing' that slipped through the filters. Whatever these viruses were, they were so small that they couldn't be seen, couldn't be cultured, couldn't be filtered.

So they didn't know in 1918 that influenza was a virus; they thought it was a bacterial disease. But clinically they knew what influenza was. When flu hits a community in the winter months it has a very distinct course through the population. It's extremely rapid: within a couple of weeks lots of people get sick, and it goes away equally rapidly. Even when you get 'ordinary' flu, it's quite a serious illness. People misuse the term all the time: they say they have the flu when really they have colds or other upper-respiratory infections caused by dozens of viruses or other agents. But when you have influenza you have extremely high fever, massive muscle aches and pains. It's not the kind of thing you take some Tylenol for and get up and go to work. You are flat out in bed for 10 days; you feel miserable. It's a really awful disease, but most people recover.

What happened in 1918 was that lots and lots of people got the flu, as *ordinary* flu, but the amazing thing was that enormous numbers ended up having devastating viral pneumonia, and secondary bacterial pneumonias, and died. And what was absolutely mystifying was that the people who had the highest chance of dying were young healthy adults, people in the 15- to 35-year age range. This is still the huge mystery of the 1918 flu: why it had this enormous impact on the healthiest, most robust segment of the population. It had a devastating impact on the military, because that's exactly the age range of people who were fighting in World War I. The peak of the pandemic was autumn 1918, and actually I'm sure it was a key factor in having the war end in November.

Troops in training camps, or in trenches, or on troop ships are always highly susceptible to respiratory diseases anyway, and here was a virus that spread extremely well but also had a high propensity to kill people exactly that age, so its impact on the military was enormous. The US entered World War I very late compared to the European combatant countries, so there were far fewer US military casualties – about 100,000 in total in World War I. But, of those, over 40,000 died of influenza. So 40% of those young, healthy, well-fed, strapping American GIs dropped dead of flu in 1918.

Anyway, it turned out the AFIP had autopsies on about 100 such cases. The retrieval system was so excellent that we were able to do key word searches for 'influenza pneumonia', find the cases and retrieve them. The cases showed up on my desk not having been touched in 80 years! Then we tried to develop techniques to see if we could tease out tiny fragments of the virus's genome from these tissues. It took about a year before we found a positive case, or worked out a technique that would allow us to do that. We just kept getting negative results. We came close to thinking maybe we should abandon the project, that it just wasn't going to work and wasn't worth the effort. But it was such a keenly interesting project that we wanted to keep going. And the more we read about this virus, and the outbreak and the devastating impact it had, the more committed we became to get the project to work. And so it ended up finally working – we found one positive case out of about 70 that we had looked at.

Initially, I was struck by the way that case looked under the microscope. It was a soldier who had developed influenza pneumonia and been hospitalised. Clinically it was recognised that he had developed a pneumonia of his left lung and the right lung was seemingly normal. Now it's quite common to have pneumonia of just one lung, and that was confirmed by the autopsy – he had massive bacterial pneumonia of his left lung that was fatal, and his right lung was almost completely normal. But what was there if you looked at the sections carefully – and it was not actually noted at the time of autopsy – were tiny areas of very acute inflammation around the terminal bronchioles in that lung that were characteristic of very early phases of the influenza viral replication.

What I think happened was that there was an asynchrony in the course of disease in that he got influenza and then had a bacterial pneumonia in his left lung, which overtook the viral infection and killed him. But the influenza virus infection of his right lung was somehow delayed by several days, so when he died it left a snapshot of the very earliest stages of the virus infecting the lung. It was a very subtle change, and it took me a while to look at enough autopsies to get a sense of that. Influenza does not have characteristic changes that allow you to be confident, to just look under the microscope and say, 'This is definitely influenza.' You can suspect it, but you can't be certain. But there was something about that case that just struck me as an excellent example, so once I'd identified it, we extracted genetic material, did our test and 'Boom!' – we found influenza virus RNA.

And how excited were you when you found it?

Very! [*we laugh*] Very, very excited. You know, nobody knew anything about that virus. The amount of material we had was vanishingly tiny, and the quality of the genetic material, the virus, was horrible. So it was a rather daunting task to go from the first part of the project, which was, 'Can we find a positive case?' to the sudden and horrible realisation, 'Damn, we've found a positive case! Now what do we do?'

We had to try to do large-scale genetic sequencing of the virus to work out what it was. I was worried that this case was not adequate for that; there was just not enough material. So we put a moratorium on using that material to sequence more of the virus while we looked for additional cases. I found another case of a person who had died, eerily, on the very same day in September 1918, but in a different army camp, that also ended up being positive. So now we had two cases. And then I got a reprieve from an unexpected source, another pathologist, Johan Hultin, a really amazing man who's in his eighties now, in San Francisco.

Johan Hultin was a Swedish immigrant, who was at med school in Stockholm and had come to the United States to do a PhD. He ended up in the University of Iowa in the microbiology department, and one amazing thing led to another for him. A person had visited the lab – this was about 1949 or 1950 – and given a seminar on influenza. He'd mentioned the importance of the 1918 outbreak and said that unfortunately the only way we could ever find out anything about the virus would be to recover it from someone who had died of flu and then been frozen. Light bulbs went off in Johan's mind. He discovered there were a number of small Inuit villages in Alaska that had suffered devastating outbreaks of this flu, and where the victims were then buried in permafrost, in the ice. So he went up to Alaska, talked to people, and ended up proposing a project in which they would exhume the bodies frozen in the ice of people who'd died of influenza and try to recover the virus from that. This they did in 1951, and they brought back lung tissue to the laboratory and tried to recover live virus.

Influenza viruses had been discovered in the 1930s, so people knew how to culture them. Electron microscopes, of course, had been invented, so they could see them. But Hultin and his colleagues were unable to recover the virus. It makes sense: the virus is very fragile, and 'permafrost' is actually a misnomer. There are continual cycles where the temperature goes just above and just below freezing, and, with biological material, freezing and thawing is the very worst thing you can do. Ice crystals form and poke holes in membranes of cells, and it causes all sorts of damage. So, basically, nothing biological survives freezing and thawing.

We published our initial findings, just tiny fragments of the sequence of the virus from our pathology material from the AFIP, in *Science* magazine in 1997. And Johan Hultin, who was in his seventies, read the article and wrote me a letter explaining this whole story.

Did you get to look at the old material that he'd collected?

He didn't have it! Here's the amazing thing: he worked with a microbiologist and a pathologist; the frozen material they had in the 1950s was kept in their freezers for a while, but then, as they could not find the virus, the material was thought to be useless and was thrown out. There was formalin-fixed pathology material made from it, which was kept till the early 1990s, when it, too, was disposed of – just a couple of years before we started this project. Really very bad timing!

So Johan proposed to go back to Alaska and get more frozen material that we could now do molecular analysis on. (Of course, no molecular analysis was possible in 1951 – the structure of DNA wasn't even determined till 1953.) He funded the expedition entirely himself. He went back and some people in that community still remembered him 45 years later. He got permission to do an exhumation; he sent us material, and it became obvious within about a week of receiving it that it was going to be positive. Johan had lung tissue from four cases: three of them were negative and one was really positive, and we had lots and lots of material. So now we had three positive cases.

So, was it thrilling when you got the stuff from Hultin?

It was absolutely, amazingly thrilling!

We ended up sequencing the main gene of the virus from all three cases. And what we found, amazingly, was that they were basically identical, one to the other, so there was no question that this was the virus. And using the frozen material it became possible to sequence the entire genome of the virus. But it took an enormous amount of effort, from 1997 through early 2005.

One of the things we concluded was that this was a bird flu virus that had adapted somehow to humans. And by coincidence our publication – the first publication of the 1918 virus – occurred in March 1997 at the exact moment that a three-year-old child in Hong Kong got infected with H5N1 bird flu virus and died. This was the first evidence that a bird virus could actually infect a human and cause disease. Before that it was thought that a bird virus infected birds and just couldn't infect humans. In that bird flu outbreak in Hong Kong, 18 people got infected, six of whom died, and so there was an enormous upsurge of interest in influenza. People were worried that the H5N1 virus was paralleling what 1918 did – causing high mortality, in young people. And so the two stories – the 1918 and the H5N1 bird flu stories – have intertwined over recent years.

While all this was going on I was still running a molecular diagnostics laboratory which was receiving upwards of 20,000 cases a year of different kinds of things. I also had administrative duties, and I had various research projects. It all got impossibly too much, so the research projects had to go one by one, until flu just rudely pushed the others off the plate and took over my life!

Okay, so you'd sequenced the virus genome, got this great string of letters – then what? What can you read from the sequence?

The answer is that we have extremely little understanding of what that string of letters means. A virus is nothing but a package of genes inside some proteins, so whether it's even alive is debatable. It's either a kind of complex chemical, or a very simple life form. Personally I think of viruses as alive because – and I'm being very anthropomorphic here – they are clever little beasts! I definitely think of viruses as living and as my adversaries.

So, we have the sequence now. We know that this string of letters produced a virus that killed 40 million people. Somewhere in there are changes that make that virus different from other influenza viruses that don't do that, so there are two fundamental questions to ask about 1918. First, what changes are necessary – what mutations have to occur – for a bird virus to become transmissible in humans? And not just able to infect one individual human, but become transmissible from one person to another, which is necessary for it to cause an outbreak? The current bird flu virus infects individuals, so if you are exposed to chickens that have the virus, you can get infected, but you cannot transmit the virus to your brother . . . luckily! So the saving grace for us, with this highly lethal virus now, is that it hasn't 'learned' to be transmissible, whatever that means. But 1918 did.

The second question is: why did 1918 kill so many people? And why did it specifically kill young adults?

That's the big question at the moment, is it?

That's the big question. And what we think is that people older than about 35 or so had some immunity to the 1918 virus that protected them, which means that there must have been an influenza virus of a similar composition circulating in the mid-1800s. That could account for why mortality was unusually low among the elderly – but it doesn't account for why there was such low mortality among children aged 5–14 years. And one possibility being explored is that the 1918 virus induced a very strong inflammatory response, and that people whose immune systems were the most robust (i.e. young adults) actually did the worst. So it's not the virus that kills you but your own body's immune response against the virus, which is so vigorous that your lungs are secondarily damaged.

It's an interesting hypothesis and there are some animal data to support it. But to me it doesn't fully explain the picture. It's still mysterious, and I'm working hard to try to figure it out. But it's a difficult problem, because we have the virus, but we do not have blood samples from people of these different age groups collected just before 1918, so we don't know what their antibody status was like [we can't analyse their immune systems].

So, unfortunately, after all this work – after more than 10 years of my life, and others' too, to find the virus, sequence it, 'make' it through the wizardry of modern molecular biology (by a technique called 'reverse genetics') and study models – we still can't explain the most fundamental question of why 1918 behaved the way it did.

I don't want to admit defeat, but it could well be that, even having the virus, having the ability to model it in animals, in the absence of this other information about infected people's immune status, we might never be able to explain this – which would be very frustrating! [*we laugh*] I would like a nice coherent picture with a happy ending, but that's still a fundamental problem.

Fascinating! So it's not just the virus itself – there are all kinds of other things in the picture, variables, that you have to investigate now?

Right. And I think this is actually a more realistic approach, because infection is really a dance between the host and the pathogen. Different people react differently to the same infection, and it's the *person* that varies.

You mentioned that 1918 was also a bird flu – how d'you know that?

Well, because it looks very much like a bird flu in its genetic sequences. We've identified a series of about 25 gene mutations that we think distinguish 1918 from typical bird viruses. Basically those are the *only* changes that distinguish it from bird viruses.

So you can see where it jumped the species?

Yes. But we don't know which of those changes are crucial for this process – whether it's all of them or some subset. Now we have to do the kind of test where we take the 1918 virus and make those mutations to turn it back to the typical bird sequence, and then see if it no longer causes infection in the lab animal. Those are exactly the kind of experiments, using very careful molecular virology work, that we're proposing to do once our new BSL3 high containment lab is functional. Some of the H5N1 viruses have a few of the same mutations found in 1918, suggesting that they may be undergoing similar pressures to adapt to humans, but they've not yet acquired enough of these changes to make them go, and one of our goals is to work that out.

So how do you feel about the project at the moment? You've obviously done fantastic groundwork, but do you feel it might actually hit the buffers at some point because there are too many things you can't test?

[*Pauses*] What I learn, the more I think about the whole influenza problem, is that it's just a very complex system, and the ideas and models that have been developed over the last 50 or 60 years to explain how it works are too simplistic, they just don't fit the complex data that we have now. So we're trying to take a new look at all of this: how viruses move around in different species; how they evolve in bird hosts; and how they evolve in humans. It's just an unbelievably complicated and dynamic system.

Tell me, as somebody who has gone very deep into pure science research, what are the advantages of being a pathologist as well?

I think it's a huge advantage. People who have training in one discipline of basic science often have a rather narrow set of experiences, and training as a physician in general – but specifically as a pathologist – I think gives one a much broader, a holistic, view of things. I spend most of my time thinking about nucleic acid sequences – those strings of letters you mentioned – but I'm constantly trying to put it in the context of the epidemiology (the occurrence and distribution of the disease in the population), of the clinical picture and so on, and I think that's really valuable.

So you see yourself as a physician, do you?

I do. And I'm very glad that I'm a pathologist. I have an anecdote to tell you that just came to mind: Tim O'Leary, my former boss, has several children, and one of his sons came into the office one day – he was seven or eight years old at the time, a very smart little boy – and I was chatting with him for a few minutes. Somehow the fact came up that, like his father, I was a pathologist, so I asked him did he know what a pathologist was. He gave me this answer, which I think is just fabulous: 'A pathologist is the kind of doctor that figures out where the problem is but then can't do anything about it'! [*we laugh*] I thought that was an excellent definition of pathology . . . Here we are, more than 90 years later, trying to figure out what the problem was with 1918, but we can't do anything about it!

The other big question is: what do you do outside of all this? If you have any time . . . !

Well, what I really want to be when I grow up is a composer, actually! Music has been the other key passion of my life, but it seemed more practical to have music as a hobby and science as a profession than the other way around. It would be kind of hard to be in an orchestra and say, 'Can I borrow your lab for the weekend?'

Since my teenage years I've been very interested in composition. As a child, I would get a score to Tchaikovsky's Fifth Symphony, put it on the stereo and think: now, how did he make that sound? And having played in bands and orchestras and chamber groups for years, I've got a sense of things, and I've read books on orchestration and am foolhardy enough just to sit and write music.

So you were trying to tease out the genetic code of Tchaikovsky's music, were you?

Yeah, I guess so. Actually, there's a poster here [*he indicates his office wall*] – I had a concert at the AFIP, where they played one of my string quartets in relation to Alexander Borodin, who is a hero of mine. Borodin was a physician who did organic chemistry research and composed on the side as a hobby. There are descriptions of how he would come out of his lab busily scribbling his musical scores, and then he'd go back to his lab, where he had beakers bubbling away. So he's a hero of mine – but he wrote more music than I have! I'm married, and I have two young children at home who are now 10 and 12, so obviously I have almost no time available for music. I hope to do more later, when things settle down in my life.

One final question: how has delving so deeply into the nuts and bolts of life affected you philosophically? Has it been 'just science' or has it influenced your views about the meaning of life?

Well, that's a very interesting question. I'm fully aware of how complicated things are, especially in the biological systems, but I guess I have to say I believe there's a molecular basis for everything we see, and it could actually be worked out. So I have *faith*, and that's kind of what keeps me going: it's faith that eventually science will be able to explain a lot of things that will have practical applications in medicine, advances in therapy, and so on. So yes, I do strongly believe there is a definitive programmatic basis for everything, and that we could work it out someday.

WHO WAS THE FIRST PATHOLOGIST?

Juan Rosai

*Director, Center for Pathology Consultations, Centro
Diagnostico Italiano, Milan*

Juan Rosai was born in Italy as World War II began and emigrated with his family to Argentina at the age of eight. His father, a plumber, was in charge of the heating system at Harrods' Buenos Aires branch. Rosai went into medicine partly as a result of parental ambition – 'You know, the dream of every immigrant is for his sons and daughters to be doctors' – and partly from having seen medicine in action when he broke his leg badly falling off his bike as a teenager. Having qualified as a doctor, he left for the United States when it seemed impossible in Argentina's political climate to practise the kind of medicine in which he believed.

In the US Rosai quickly established himself as a first-class diagnostician, specialising mainly in tumour pathology. Over his long career, he has identified many new disease entities, one of which, Rosai–Dorfman disease, is named after him and his fellow researcher. Rosai is also an acclaimed teacher and a prolific writer. He is author of the classic textbook, *Rosai and Ackerman's Surgical Pathology*, which he took over from his mentor, Lauren Ackerman. He has written also a book describing the 'very rich' history of surgical pathology. He says, 'The founders of surgical pathology were fading out, so I thought maybe somebody should record their experiences before it was too late . . . I think we have an obligation to know where we came from and why we do things the way we do.'

My choice to become a pathologist came very early. Pathology was introduced in the third year of medical school, and I fell under the influence of a very charismatic professor, Eduardo Lascano. We were close to 4,000 students in the first year, so it was impossible to have any sort of relationship with the teachers. But some professors were allowed to have special courses for small groups, and Dr Lascano happened to have one in pathology for about 20 students. He convinced me that pathology was the scientific basis of medicine. So I was already decided by the fourth year that I was going to be a pathologist.

When you chose pathology were you frustrated at the idea that you wouldn't be working directly with patients?

No. Dr Lascano and others convinced me that the excitement I would get from the scientific aspects of medicine would compensate for the lack of direct patient contact – which, incidentally, is not completely true; we interact with patients quite a bit.

The day after I graduated from medical school, I started my pathology residency. And I would have stayed in Argentina had it not been for some political upheavals in the country and in the hospital where I was working.

It was an inspiring and ultimately sad experience, because I have never worked with a more idealistic group of people in my life. Medicine in Argentina at that time was divided between the public health system, which was grossly under-funded, and the private clinics. For most physicians this was a good arrangement, because they made a good deal of money in the clinics. Only the people who could not afford otherwise ended up in the hospitals.

Well, just as I was graduating, a group of Argentinian physicians who had trained in the States returned and decided, 'Let's try to change this rotten system. Let's see whether we can build something like we saw in the States – a modern hospital.' They went to see the Minister of Public Health, who liked the idea, but said, 'There is no chance to implement this in hospitals that are already running; they will not let you. But it just happens there is a hospital in Mar del Plata.' This hospital had been built by Perón during his first presidency. It was just finished when the revolution toppled Perón and so it remained closed for something like 10 years. He said, 'If you want the hospital, you can have it. Do whatever you want; we will support you if we can.'

It was really an adventure. We left our families and friends, and went to live in that hospital like monks. I was the only resident in pathology, and I handled every biopsy, every surgical specimen and every autopsy under the supervision of Dr Lascano, who had also decided to participate in this adventure. And it worked very well for four years; it looked like it was going to survive despite fierce opposition from the private clinics, which were in competition with us. They did not want the city to have a public hospital where good medicine was practised. They achieved their end when there were national elections, and a change of the government. The new Minister of Public Health worked in a private clinic himself, so he said, 'Well, this is a very interesting experiment, but four years have passed; the time has come to evaluate it.'

It seemed reasonable enough; the only problem was that he sent us an evaluator who was the son of a director of a private clinic, and who had graduated only a month before. It was too much. When he came for the evaluation, we were all there outside the hospital entrance, and we didn't let him in. Next day he came with the police, and we resigned *en masse*, every single employee of the hospital. Some people went back to the cities they had come from; others stayed and tried to do privately something similar to what we had done before. And others, like myself, decided to go to the States.

So you left Argentina . . .

I was very lucky: the week that I had decided to go, Dr Ackerman gave a talk in Buenos Aires. Lauren Ackerman, from Washington University, St Louis, Missouri, was at the time one of the most famous surgical pathologists in the world. He was the author of *the* textbook of surgical pathology. I went to see him, told him I was interested in going to the States, and he took me on as a resident. I stayed in St Louis as a fellow and then on the staff, and I remained in the States for 35 years.

Ackerman's book was the 'bible' of surgical pathology. It went through several editions, and at the time of the fifth edition he asked me to take over, and I have done it ever since. I started recently working on the tenth edition.

So when you started in St Louis did you see your future as a surgical pathologist doing the service work, or had you got research interests?

Both. At the time the research was anatomically based, in that it was performed using the same tools that we used for our diagnostic work. In a way our research was an extension of our service work: it was, as we call it, clinico-pathologic research, and Dr Ackerman was a master of that school, which is interested not so much in the basic mechanisms of disease, but rather in the clinical implications of the pathologic findings. So after having trained with Dr Lascano in Argentina and Dr Ackerman in the States, I became that kind of surgical pathologist. I am not, and have never been, a basic scientist studying the *mechanisms* of disease.

So what were you seeing in your service work, and what interested you most?

Well, surgical pathology basically refers to the study of any tissue that is obtained from the living patient, as opposed to tissue obtained at autopsy. I'm biased, but perhaps the most important branch of surgical pathology is the one that deals with cancer and tumour-like conditions. So my speciality, if I must call it such, is oncologic pathology.

I find it the most interesting from many angles, and I think cancer is where we pathologists contribute most. So-called medical pathology – the study of liver diseases, renal diseases and so forth – certainly contributes a lot. But the pathology in these cases is just one of many pieces of information that the clinician uses to decide what disease it is, how to treat it and what the prognosis will be. In oncology what we do is absolutely essential: despite all the advances in molecular biology and other disciplines, the diagnosis of solid tumours is still based – in the overwhelming majority of cases – on what we see under the microscope.

Some people say that all the new technology is going to make pathology increasingly redundant. Do you think that's so, or is pathology still the gold standard?

You know, I have heard this now for 35 years. There is always someone saying that tomorrow they will describe a DNA test that will put pathologists out of business. It hasn't happened, and I don't think it's going to any time soon. Of course, the new tests are improving our capacity to categorise diseases, and in particular fields like haematopathology they are becoming very important. But for most solid tumours, microscopic examination remains essential – and often the only thing you need.

So with a really good eye you can see the differences, can you, even if there is a genetic basis to the differences?

That's right, because morphology is like a grand summary of all the genetic events in a cell. I mean, any genetic event that is biologically significant will translate into some morphologic change that you should be able to see under the microscope. One example I like to give is of a young woman with enlarged lymph nodes in the neck, which are biopsied. The pathologist recognises the type of tumour, and he can tell the surgeon, 'This woman has a metastatic tumour, it's glandular, made up of thyroid follicular cells, so the primary is in the thyroid on the same side as the lymph node. The chances are that it's multicentric, it will pick up radioactive iodine and the patient will be cured of the disease.' That's a lot of information from just a simple H&E slide – and very cheap to obtain.

And genetics wouldn't have told you any more than that?

Not even as much.

So what was the state of knowledge about tumours when you started pathology?

The knowledge of tumours had advanced quite a bit, but it was still possible to discover new diseases and new findings with the light microscope. In fact, there is a disease named after me, Rosai–Dorfman disease, and it was discovered just looking down the microscope.

Ronald Dorfman is a South African pathologist who had also come to work with Dr Ackerman. In one of his conferences for residents he said that a particular disease term that pathologists were using, reticulo-endotheliosis, was, in his opinion, a 'waste basket' diagnosis. He said, 'I'm sure if we were to look back at the cases we're calling reticulo-endotheliosis we'd find they are all kinds of things. It would be an interesting project for somebody.' After the conference I said, 'I'd like to do this project with you.' He agreed, and I started pulling out from the files all the cases that had been diagnosed as reticulo-endotheliosis. And he was right: there were all kinds of diseases in there.

But the most interesting part was that, in the review of those several hundred cases, I found two that had an appearance that was totally different from the others, identical to each other, and very, very peculiar. Reticulo-endotheliosis is supposed to be a malignant disease. Both of these cases had been called malignant, but I looked at the follow-up and found that both patients were alive and well and perfectly cured without treatment.

So I mentioned this to Dr Dorfman, who looked at them and said, 'Yah, not only do they look funny, but they look identical to two cases I have from South Africa.' He found those slides, and they looked the same. And those patients were cured too. So we had four cases; it was enough to suggest that it was something distinct, 'an entity' as we call it. So in 1969 we wrote a paper on the four cases, and we got further proof that it was an entity because pathologists reading the paper and looking at the pictures realised they had seen something similar themselves, and they started sending us their cases. A few years later we were able to write a second paper with 34 cases, and now there are hundreds.

But why had pathologists been prepared to throw this into the 'waste basket' of diagnoses if it looked so different?

Well, you know what they say: 'You only see what you know.' The tendency of most of us is to try to put what we see into one of the boxes that have already been created. And if you cannot do that you either force it into one, reasoning that it must be an atypical case of a known disease, or you say, 'I'm not good enough to recognise this.' That's what happens with most entities, and if you look back and ask, 'How come something so obvious was not recognised before?' that's the reason. But once people are familiar with it, they recognise it again easily. Today a first-year resident will recognise Rosai–Dorfman disease at a glance.

What exactly is it and where does it appear in the body?

We called it 'sinus histiocytosis with massive lymph-adenopathy' at first. Later somebody suggested that we call it Rosai–Dorfman, which is actually not quite right because it turned out there was a French pathologist who described some cases before in an obscure journal of tropical medicine. In the most typical case it will affect a young person, often of the black race, with bilateral, pretty massive lymphadenopathy [swelling of the lymph nodes] of the neck. I mean so massive that the clinical diagnosis is usually that it is something malignant: a lymphoma or leukaemia. But if you recognise it and do nothing, in most cases the disease will regress of its own accord. There is no specific treatment for it; but fortunately in most cases you don't need it. All you need is to recognise the disease and do no harm.

And what did finding this disease teach you specifically?

That there was still a lot to be learned from morphology. I remember when I did this work another fellow said, 'Well, you were lucky, because this is probably the last entity that is ever going to be described.' And it was not true. *Lots* of additional tumour entities have been described. I see many cases in consultation, which by their nature tend to be unusual, and sometimes one hits you because it's maybe something you've never seen before, and it doesn't 'fit'. So you store it in your memory. Then if you see another case you put them together in your mind, and if you see a third case you start thinking: maybe this is something. Then you start your research to prove it.

Let me give you a good example. Many years ago, in Minnesota, I got a case in consultation of a young woman who had delivered a baby, and a few weeks later, during the check-up, they found a little polypoid lesion in the uterine cervix. They took it out and it looked under the microscope like a malignant tumour, a sarcoma, and that's what it was called. I thought it strange that a sarcoma would appear so quickly after a normal delivery. Six months later another young woman had the same thing – and another biopsy was called sarcoma. I thought, no, there's something not quite right. Two months later a man had a TUR [transurethral resection] of the prostate for benign disease, and a few weeks later he developed a little polyp [growth] and it was biopsied and diagnosed as sarcoma. Then I said, 'No, this cannot be . . .'

So I went to the files and looked for other cases that had been called sarcomas either in the prostate or in the cervix, in which the patient had had an operation in the area a few weeks before. I found something like seven or eight cases, and on follow-up they were all fine, so I concluded that it was not a sarcoma. It *looked* like a sarcoma but it was, in fact, an exuberant reaction to the operative trauma.

They were what we call 'pseudosarcomas': benign reactive changes. Problem was that they'd been called sarcomas: the patient with the prostate had, as a result of the diagnosis, had a total prostatectomy.

But what did it take for you to put all the pieces together in those cases?

Well, visual memory is very important.

And is that a natural endowment, or something you can learn?

It's both. Like tennis, you can certainly improve by taking lessons, but if there is no inborn talent, you are never going to be first rate. I think the same is true of surgical pathology. You can improve over the years, but there is no question that some people have a better 'eye' than others. You also need imagination – and *conviction*. That's one of the things I try to teach people: 'If you see something that doesn't fit any disease you know or what is in the books, don't assume that you don't recognise it because you are no good. Maybe it is something that hasn't been described yet.'

Besides the pleasure of discovering new things, what are some of the other rewarding aspects of your work?

Well, another big event for me was when Dr Ackerman gave me the book to do. There are very few pathology departments in which you would not find that book. To think that through it I am influencing, or helping, pathologists all over the world is a great source of pride and satisfaction.

So what are your main aims when you sit down to write it?

Basically to help people whose job is to make microscopic diagnoses. I try to be very practical in the sense of pointing out the pitfalls to a diagnosis and listing the clues to its recognition, as well as providing an updated terminology and classification of the various tumours. It's not a book for medical students to learn pathology. It's a book to be used where pathologists are working, as a reference when they need it.

Give me some examples of tricky diagnostic situations where it might help.

Often the most difficult are not the very rare diseases, but the diseases we all know, and where you are trying to decide, when it is a borderline lesion, whether to call it benign or malignant; carcinoma *in situ* or invasive.

A very difficult thing in surgical pathology is to express in words the mental mechanisms that you use in order to reach a diagnosis, because the process works to some extent at a subconscious level. People will show you a case, you realise immediately that it is an alveolar rhabdomyosarcoma, so you make that diagnosis. But if they ask you, 'Why did you say that?' then you have to think! You go ahead and list the reasons, but you haven't thought of them, at least consciously, while you were looking. It was just an instantaneous recognition.

I remember some years ago, when I was giving sessions in which pathologists came from other hospitals to show me their difficult cases and we gathered round a multi-headed microscope, there was one pathologist who taped all the sessions, trying to figure out the mental mechanism by which I went about diagnosis. And he abandoned it; it was too difficult!

As well as the textbook, you have also written a book about the history of pathology – can you tell me a bit about that?

It's called *Guiding the Surgeon's Hand*. The history of surgical pathology is very rich. Nothing had been written in a comprehensive fashion, and the founders of surgical pathology were fading out, so I thought maybe somebody should record their experiences before it was too late. Also I think we pathologists have an obligation to know where we came from. Because many of the things we do now, we do because somebody before us instituted things that way: knowing how things were decided helps you to understand why you do them and maybe to modify them.

Pathology as such started with the autopsy. It began during the Renaissance in Italy, and it worked in the following way: a physician would follow a patient and if that patient died, the physician would perform an autopsy to try to understand the reasons for the symptoms and signs. Most people did not record their experiences, or the books were lost. But it so happened that one of them, Antonio Benivieni, was very good at describing his experiences, both the clinical and the pathological, of autopsies, and after his death his brother found his record book and had it published in 1507. It is recognised as the first book in anatomic pathology.

So these were just clinicians who cut up the bodies of their patients to find out more about their illness?

That's right. But during the Renaissance performing autopsies was the rule. People think that autopsies were only started in the 1800s, that for religious reasons they were not allowed before that. But they were so common in Florence during the Renaissance that Benivieni, in his book, comments on a case in which the autopsy was refused. He was indignant. He said, 'It was refused because of what superstition, I don't know.'

Where were autopsies performed?

Usually in churches. The funeral would be in the church and then the autopsy would be done in the basement. Actually, that's where Michelangelo and Leonardo da Vinci did many of their anatomical dissections.

The physician needed the consent of the family except in cases of suspicious deaths, or epidemics, like today. When they had epidemics of bubonic plague, for instance, they needed to be sure that someone had died of plague and not something else, and therefore the family could not refuse. As I say, autopsy performance was widespread.

So these doctors were pioneers really?

That's right. And actually those were true clinico-pathologic studies, because the physician who had done the autopsy would write to the patient's relatives explaining what he had found and giving them advice and so forth. So the approach was very clinically orientated – it was trying to understand diseases from the symptoms and signs. From Italy the discipline moved to England, with pretty much the same approach. From England it moved to France – now we are in the years 1600 and 1700 – and from France it went to Germany in the 1800s, where it took a very significant turn under the influence of Rudolf Virchow.

Virchow was a scientist. He was not very interested in diagnostic pathology; he was interested in understanding the mechanisms of disease. Almost single-handedly he transformed pathology into a science – into the scientific basis of medicine. And when pathology was imported into the United States in the late 1800s, it came through the German schools, and therefore it replicated the German philosophy. The departments of pathology in all the major medical schools were research departments. This was where most of the medical research was carried out – to the point that the journal they published was called the *Journal of Experimental Medicine*.

Meanwhile, there were surgeons, gynaecologists, dermatologists and other physicians who were interested in learning the pathology of their patients in order to treat them better. They would go to the pathologists but found them not interested: they were busy doing science. So these physicians said, 'Well, okay, if you won't do the pathology for us, we'll do it ourselves; we'll become pathologists.' And so what happened in the States in the early 1900s was that departments of surgery, gynaecology, dermatology and other specialties developed their own pathology laboratory within their departments. The surgeons were most active in this endeavour. That's why it's called 'surgical pathology': because it's pathology done of surgical specimens, by surgeons in labs within the surgical department. The chairman of a clinical department would appoint somebody from his own department to be the pathologist.

So the first surgical pathologists were surgeons, not scientists. In fact, they were belittled by the members of the pathology department, who regarded them as 'practitioners', whereas they saw themselves as scientists.

In a way, surgical pathology today is a continuation of the original line of pathology that goes from Italy to England to France – the practical, clinical-based pathology. And what's happened over the years is that there has been a merging of the two branches, so that beginning in the 1960s the pathology laboratories housed within clinical departments moved to the pathology department. One reason was that pathology had become too complex a discipline, and it was very costly to duplicate those facilities for each clinical department. It made sense to gather all under one roof. But the philosophical gap between the two has remained to this day.

So as a young medical student, are you essentially learning surgical pathology or 'scientific' pathology, or can you go either way?

Well, that's a very good question. And in a sense that has been one of the big conflicts, because you are talking about two different approaches. Theoretically both are being taught; in practice it is mainly the science. It's called pathology and it's the mechanisms of diseases, and there are books that deal specifically with that. Surgical pathology is taught together with the clinical specialties. For instance, if you take lung diseases, there will be among the teaching staff a surgical pathologist who will teach the medical students lung pathology from a clinical standpoint.

So when you decided to do pathology, were you aware of the split and did you know which direction you wanted to follow?

I gradually became aware of it, and actually Dr Lascano, my teacher in Argentina, was, I realised later, torn between these two approaches; he wanted to do both. I think he wanted to be more of a scientist, but because of various limitations, he did mainly surgical pathology. But Dr Ackerman was on the surgical pathology side without question. Actually he would make fun of the scientists: he called them 'the mouse pathologists'. Once I heard him say, 'The scientists always tell you they study diseases in animals with the purpose of applying the findings to humans. I do the opposite. I work in humans, but if they want to apply those findings to mice, I have no objection whatsoever!'

I hadn't fully appreciated these distinctions, though people have talked about the service side and the research side. But there are people who seem to be doing both.

It's the most difficult thing in the world. Many people are actually caught in this dilemma, and some people, I guess including myself, suffer from it. But at some point in your career you have to make a choice. I made the choice by going with Dr Ackerman, who was a leader in the field of surgical pathology.

Is it a mindset that's different?

Yes, it is a mindset. I have seen some people who seem able to do both; my experience is that they are the exception. And if you look closely, you will find that most are very good at one thing and just manage on the other. But very good at both? Very, very few. I am a surgical pathologist, no question about it.

And what is it that most appeals to you about that?

Well, first of all the satisfaction of looking at cases and making a diagnosis – the feeling that I'm helping patients. And the fact is that I do not have the background or the time to do science. But by looking at the cases I can come up with ideas that I can transmit to others who have the means to explore them.

So you have raised questions for the pure scientists, have you?

Actually that's something I have tried to do in every place I've been – I have to say, generally with meagre results. Most basic scientists don't want to hear from us. They don't think we have anything to tell them. They feel as a group that the only way to really understand the basic mechanisms of biology or diseases is through basic science – and I think there may be some truth in that. With plain observations maybe you can get some ideas, but it will never get too far.

They also think – and I hate to say that maybe they are right – that as a group they are more intelligent, more inspiring, more imaginative people than those on the surgical pathology side! And actually, I'm like Lascano: I'm frustrated, because, although I do surgical pathology, and I'm very satisfied with doing this, I lament the fact that I was not able to do 'pure' pathology. I wonder what would have happened to me if, instead of taking this route, I had taken the other? I don't know.

What would you say is your motivation for the work that you do?

That I like what I do: I like thinking about the diseases; I like teaching young people to become good pathologists; and I like to help patients by making the right diagnosis.

I have always liked the idea of contacting patients directly whenever it was indicated. Not only do they appreciate it, but sometimes you get information that you have not gathered from the clinician that may influence your diagnosis a great deal. But I have found that on the whole clinicians do not welcome that. In fact, I was criticised more than once in the States – a surgeon saying, 'Why did you call my patient? If you want to know something about that patient, call me.'

Here in Italy my consultation practice is very different from the one in the States. The cases I get from the States are almost all from pathologists. The consultations I get here are almost always directly from patients, often because the oncologist advises them to consult me. And it's a very personal encounter. The patients bring the specimen and ask to talk to me. At the beginning I was a little nervous, but I have found it a very satisfying experience. First of all, the patients appreciate it tremendously. Many have said that nobody has given them time and explained their disease in the way I have. And second, again, they provide information that the clinician may not think of giving me that is important in the interpretation of the slides, such as that other members of their family have the same problem.

Okay, as someone who gets the difficult cases, tell me about some of the most memorable you have seen.

Well, one is the group of cases out of which I describe an entity for the first time. Second would be the cases in which a wrong diagnosis has been made of some terrible sarcoma that would require a radical operation and I see the slide and realise it is a benign process and that nothing need be done. That has happened quite a few times, though not all are as dramatic as that. But, for instance, the thyroid, which is an organ of special interest to me – I think I have saved a lot of thyroids by telling the surgeons that removal of the whole gland is not necessary.

So what entities have you found besides Rosai–Dorfman?

One is called desmoplastic small cell tumour. It is a malignant tumour that usually involves the peritoneal cavity of children or adolescents, usually males, and it's a very aggressive, often fatal, disease. That has been very gratifying to me for another reason. I identified the entity on purely morphological grounds. I wrote a paper, and an argument soon arose as to whether it was really an entity or not. Meanwhile, somebody described in those patients a specific chromosomal translocation – many tumours are associated with specific chromosomal translocations – which led to the discovery of a specific gene fusion resulting from that translocation. That has led to some new diagnostic tests, and to some promising therapeutic attempts. And it all started with the morphology.

On such occasions is it just serendipitous that the two lines of enquiry come together to provide answers, or does it start from there being a clinical question in the first place?

Well, actually that's the satisfying thing about pathology. When I was discussing not too long ago the desmoplastic small cell tumour with a group of people and pointing out that without the molecular biology we would not have gotten where we are today, somebody said, very generously, 'That's true, but if you had not described the entity in the first place, we would never have known of its existence.' The pathology, the morphology, was the most original aspect of the study.

So, first you discover the entity morphologically, and then the geneticist will say, 'Oh, this is associated with this genetic event.' It almost never happens the other way round.

So tell me, when you start looking at slides, how much are you seeing a patient behind the pathology and how much are you looking at fascinating science?

Well, obviously you start looking at the slide, and you get excited about it, and actually you then feel guilty because you see a terrible cancer and you say, 'Oh, how beautiful!' and then you realise that that tumour is killing somebody. But if you are doing your job correctly, your work does not end with a diagnosis; you must make sure the surgeon gets the message about what that means for the patient. I am very direct in that I make the diagnosis and then suggest what I think is the correct therapy. Not everybody likes that. There is some controversy between surgeons and pathologists about that. I personally think it's our responsibility to provide them with general information that we have learned from previous cases of the same disease, which then they can use in any way they want.

What would you say being a pathologist means to you in terms of your identity?

It has always been very important – most of my life was consumed by pathology. Actually I feel some guilt in that regard, of not having devoted enough time to my family. I have three sons – Alberto, Carlos and John. Fortunately, all three turned out to be wonderful and successful kids. But the fact remains that I did not spend much time with them. But now I am 67, I think I have a more balanced view of life.

Was it your ambition all along to come back to Italy? Because you were a very long time in the States.

Yes, but I always felt Italian. I grew up in an Italian house; we spoke Italian, and most of my friends in Argentina were Italian. So I always felt this Italian 'thing'. I was warned over the years by many people, including my wife, Maria Luisa Carcangiu, who's also a pathologist, that the picture I had of Italy was nostalgic and unrealistic. And they were right, to some extent.

You got back and found it wasn't what you had thought?

Yes. I found that reality didn't match completely my idealised view! Still, I'm glad that I came back. I like it a lot despite some frustrations.

One thing that bothers me, though it does not affect me personally, is that many professional appointments are made more on the basis of political and social factors than on merit. It touches every aspect of professional life, and it is very difficult to create an environment of excellence if you choose people on that basis.

Tell me – in your consultation work, how often do you get cases that completely flummox you?

It happens often, and I have to write to the referring pathologist, 'I am so baffled by this case that I cannot even tell whether it is benign or malignant; I just don't know what it is.' Then I may suggest they send it to another 'expert'. To quote Dr Ackerman again, he often started a consultation letter saying, 'This is the first time I have seen a case like this. But then I have to say this every day!'

That's one of the great things about pathology: you never get bored. It almost seems like the variety of diseases is infinite.

SCIENCE IN THE SERVICE OF HUMAN RIGHTS

Derrick Pounder
Professor of Forensic Medicine, University of Dundee

Derrick Pounder was brought up in a mining community in the South Wales valleys and was the first member of his family to go to university. His background gave him a keen social consciousness, and he has devoted much of his time and expertise as a forensic pathologist to the cause of human rights, investigating torture, political killings and genocide in countries such as Bosnia, Kosovo, Turkey, Tunisia and Israel. He was a founder member of Physicians for Human Rights in the UK; a role that demanded the passion of advocacy. But as a professional expert in the field, objectivity is essential, he says. 'As a scientist you have to leave out a lot of that passion in order to ensure that . . . your conclusion is not clouded by emotion.'

Pounder's belief in the importance of serving the community extends to his routine work, too. He is always ready to explain post-mortem results to bereaved relatives, for example, or to provide evidence to support public health measures such as drink driving directives or the wearing of car seat belts. 'Sometimes we forget where [the momentum for safety laws] started,' he says. 'It started in the autopsy room. So we're in the business of investigating the dead in order to assist the living.'

Pounder became a forensic pathologist largely as a result of chance and inspirational teaching. As a schoolboy, he had always been interested in science, but he was later encouraged to study medicine at university, and that is where his career began.

———

After I got my medical degree, I still wanted to do science. Pathology is the foundation science of all medicine, so after a few years of hospital practice I decided to train as a pathologist.

I started that in Dublin, and it happened that, by pure chance, the professor of forensic medicine for the College of Surgeons in Dublin, who was also the coroner for the city, was at the hospital I was working in. We did the coronial autopsies for that part of the city. I did about 200 a year, and it was wonderful. It was fascinating – I was amazed by it. I had the pleasure of being able to discuss the case with the professor in the morning and then attend his court and give evidence before him in the afternoon. So it was a very comfortable introduction to forensic practice, both the science and the courts, and I decided that was what I wanted to do.

What aspect was particularly fascinating?

The science was fascinating . . . The problem-solving, I suppose, was the fascinating part. And also the fact that you were using physical skills, your own personal physical skills, in dissecting the body, and using your own observation – not just your eyes, but your sense of smell, your sense of touch as well – and trying to put all the pieces together with the background information to draw the conclusions. It was 'medical sleuthing', if you like, on a day-to-day basis. You never knew what you were going to get as a problem that day – whether you'd be easily able to solve it, or if you'd have to go to the books and the literature, or discuss it with colleagues. So it was really a daily challenge – and a pleasure.

What were some of the most interesting and challenging things in your very early days as a forensic pathologist?

In the early days the most challenging thing was to spot a tuberculosis case, because there was an infection problem, and we had quite a few TB cases in Dublin. I remember one case in which I'd missed the diagnosis, and it wasn't until about six weeks later that I discovered that I'd done an autopsy on a TB case. Of course, there are problems in terms of transmission – TB's a serious matter for the hospital. I also recall that in the entrance hallway of that particular hospital there was a plaque on the wall to a young doctor who'd died in the 1930s, a few years after graduating, and he'd died as a result of septicaemia from performing an autopsy. So health and safety and problems of infection were challenging. And the forensic cases, because they're medico-legal, involved accidents, suicides . . . At that time I wasn't doing any homicide cases at all. But some of the suicides – unravelling what had happened there, when people were perhaps brought out of the river having been there a while – could also be a challenge.

How do you manage the risks involved in a job where you could be dealing with dangerous diseases? Do you always 'gown up' and dress as though whatever you're dealing with might be lethal?

Forensic practice has changed enormously over the years. When I started out, the professor who was observing me doing the autopsy would wander in wearing his suit and stand right at my shoulder while I, gowned up of course, made the dissection. And in later years I did the same thing. These days that's totally unacceptable in terms of risk control. We deal, for instance, with HIV-positive cases, we deal with hepatitis B and hepatitis C cases. We even had a rabies case recently here in Dundee. We have to perform these autopsies, because we need the information – to help the living, essentially. So there are now a whole range of health and safety procedures – we're gowned up and gloved up and double-gloved, with visors in front of our faces to protect us from splashes. I don't think you see it like that on television too often – it probably ruins the scene!

Tell me, what do you feel about the general public's perception of pathologists as 'doctors of death'? How important is your job, and how important is it, say, that you did an autopsy on the man who died here of rabies?

Public interest comes and goes, depending on what happens to be in the media at any one moment. Public perception of pathologists is that we do autopsies. Well, of course, *I* do. I make a living out of autopsies. But I'm in a very small minority amongst pathologists. Most pathologists are not performing autopsies; they're in hospital practice and doing laboratory work to support the investigation of illness in the living, and then the treatment.

Those of us who do autopsies mainly do them for medico-legal reasons. These are deaths in which there's a community interest in the investigation – 'community' because, first of all, we want to investigate any possible crimes, but, secondly, because we want to prevent further deaths as well. All the public health measures – in terms of carbon monoxide poisoning, seat belts, drink driving and so on – all that derives from the investigation of fatalities. And it takes place over a long period of time. Sometimes we forget where [the momentum for safety laws] started – it started in the autopsy room. So we're in the business of investigating the dead in order to assist the living – there's no other reason for doing it.

Looking at something like seat belts – they're compulsory now in this country. But where would the pathologist have had an input into the discussion about the wearing of seat belts?

Some years ago we had many more fatalities on the road than we currently have, and at that time the documentation of the injuries that killed people was made by pathologists in the autopsy room. All of that information was gathered together, and it was analysed according to whether the person was a pedestrian, a driver, a front-seat passenger, rear-seat passenger, motorcyclist, and so on. And out of those studies came legislation concerning seat belts, crash helmets for all motorcyclists, drink driving, of course, and a whole variety of other rules. So it was the analysis of aggregated data from a large number of fatalities over a period of time. First of all it was recommendations, and only later did it become legislation.

I understand you're also involved in human rights work – how did you get involved in that?

I was always interested in human rights. I've always been sympathetic to the underdog; it comes from my cultural and social background. In the mid-1980s a few forensic practitioners who were interested in human rights found that we could use our professional skills in support of those rights – we could bring together our professional interest and our social interest, and we began to form an informal international network. We liaised with the various human rights organisations, and we started to consult for them in the same way that we would consult for lawyers nationally.

And really that's logical, because human rights law internationally is fundamentally no different from national law, in terms of protection of people from crimes against the person – torture and extra-judicial killings in a human rights context. We support the legal system nationally by providing expertise, and we support the legal system internationally by providing expertise. Of course in the mid-eighties it wasn't a very popular thing to do; it was very much criticised. These days human rights are on the international agenda, and it's much more *the* thing to do, and so we have many more colleagues now who are able to join us because it's not as damaging to their career.

So you really stuck your neck out at the beginning?

In the beginning, yes. It was difficult, because it was seen as highly political, highly contentious. Even if you were based in a country that was relatively sympathetic to the concept of human rights, there was still a feeling that it was a little beyond the pale for a professional person, who was after all mainly serving the criminal justice system, to be involved in something potentially as sensitive as these human rights issues.

Tell me about some of the cases you've been involved in, and what your task as a forensic pathologist is.

I consult for human rights organisations in the same way that I would consult for lawyers within the country. These are cases in which there's an alleged crime against the person. These are cases of torture, extra-judicial killing, or something of that kind. And the information I get might be only a photograph of a dead body or an injured person taken by relatives. It might be a medical report, or an autopsy report performed in another country. And I would evaluate that and give a written or a verbal opinion to the human rights organisation, then advise them whether there was solid medical evidence to support the allegation, and whether it was a case that they could pursue.

For example, I remember a case I did for Amnesty International – it was a Tunisian death, and the man had been arrested and then turned up dead on the roadside. The police and the pathologist said he'd been knocked down by a car, when clearly the autopsy results showed he'd been tortured to death. So I wrote an opinion to that effect, and Amnesty was able to pursue it. It was taken up by a whole variety of international agencies, and the Tunisian government was pressed and pressed again to explain themselves and found themselves in a very difficult position. So that's one aspect.

Another aspect is to actually go to the country and have a look. I've been to a whole variety of countries with a whole variety of agencies. Some of them are human rights organisations, like Amnesty or Physicians for Human Rights, and we would perhaps visit a scene of death and make an interpretation of the findings in the building – gunshot wounds in the walls, for example. I went to Turkey with two human rights organisations and some lawyers – we went to investigate the killing of some members of Dev Sol, a leftist armed group, and we found evidence in the building that people were probably executed there, just from the location of the gunshot marks. They were all on the floor in one room, and the shooter had clearly been in the same room. There had allegedly been a gunfight in the building, but there was no evidence of that. So that's a practical example of using forensic skills to derive evidence.

I've been to other countries and performed autopsies, either with a professional colleague in that country as an outside observer, or sometimes I've performed the autopsies myself. I've been to Israel to do that on several occasions, looking at deaths of people in the custody of the Shin Bet, Israel's security service. In 1995 I was involved in one notorious case where the Shin Bet shook a man to death, in the same way that you sometimes hear that adults have shaken a baby to death.

That case, although terrible, was a fascinating one from a scientific and legal and human rights perspective. The man was arrested and was brain dead in hospital within 24 hours. The authorities gave no information about how he came to die, other than that he'd collapsed during interrogation. They wouldn't even tell their own pathologists what the circumstances were. We performed the autopsy, and the physical evidence was clearly that he'd suffered a brain injury, and I went further and said that I believed that was due to shaking. We had good evidence that they did shake people.

Interestingly enough, the lawyer for the family said that he would phone the police and ask them whether they had shaken this man. I told him, 'No, don't do that. Phone the police and ask them how many *times* they'd shaken him.' So he did – of course, the conversation was in Hebrew; I couldn't understand it – and he told me afterwards that there was this stunned silence at the end of the phone and then the man asked, 'How did you know?'! It became a notorious case because it showed that such methods of interrogation could actually kill people. And ultimately it led to a ruling by the Supreme Court in Israel that interrogation by shaking was unlawful.

Have you ever had to go to places where perhaps there are mass graves, and you've had to tease out what's happened?

I've been involved in the Balkans – as, of course, have large numbers of my colleagues. One of the – I think – very commendable things concerning forensic pathologists in the UK is the amount of time they've given freely to support the investigations in places like Bosnia and Kosovo. Any of those cases are extremely difficult. There are individual graves and multiple graves and mass graves, and my experience of those has been largely assisting with the identification of the dead for humanitarian purposes.

Because, of course, performing the autopsy is not just about finding the cause of death; it's also about getting any other information of value. The information may be valuable for a prosecution in terms of physical evidence, or it may be of value in a humanitarian sense – in terms of identifying the person and returning the body to the loved ones so that they can bury it decently and complete their grieving process. I've been involved in showing clothing of the deceased to families, discussing autopsy findings with families, and trying to reconcile the information they have with the information we've discovered from the body in order to make the identification.

In Sarajevo there is a very important organisation called the International Commission on Missing Persons, and I was a member of their scientific advisory board. This is an organisation founded by the Americans, on the initiative of President Clinton, some years ago to identify the missing dead in the Balkans. It's a massive programme involving obtaining blood for DNA testing from hundreds of thousands of relatives of the dead, and testing bone samples from 20–30,000 bodies using a computer programme to match them. This is the cutting edge of science, and the wonderful thing is, it's beginning to work. We're beginning to get identifications at about the rate of 100 a month, due to the application of this technology.

How important is it to families that they do know something of what's happened – even if you can't actually bring a body back for them?

Of course, every family wants to know whether a loved one has died. Very often the circumstances are such that an objective observer would say, 'They must be dead.' But families naturally cling to hope. No one really wants to believe that a loved one is dead unless they are faced with the objective proof, and so very often there is a long period of denial.

Some families have to have the body to convince themselves that the person is dead. Indeed, in some cultures not having the body can make life extremely difficult in terms of the grieving process. Cultures vary, of course, but the Americans, for example, have made a huge effort to discover and return the bodies of their dead from the Vietnam War, and they spend large amounts of money to do that. That's part of their culture. We have a slightly different culture in Britain: we leave our war dead on the battlefield and bury them there overseas, such as in the Falklands. We don't usually repatriate them, although this has changed very recently.

So there are cultural differences, and we have to recognise those differences and serve people in a way that is appropriate to them.

I read a very moving story in a newspaper interview with you about clothing – can you tell that story?

One of the methods of identifying the dead is by the clothing they're wearing. In poorer communities the clothing may be handmade, or hand-repaired, which makes it unique, and so long as you can be assured that it hasn't been transferred from one body to another, it's good evidence of identity. So in the Balkans we've recovered the clothing of the dead and washed it. Of course, the bodies are so badly decomposed that, even if you wash it, the clothing still stinks terribly. But having done that it's then laid out – the clothing of many people, perhaps hundreds of people – in an old building, an old warehouse, anywhere that is covered and has reasonable ventilation. And then the local people are invited to come to see if they can recognise the clothing of a loved one.

I recall a man identifying clothing, which he said were the trousers of his father. But they were mass-produced, so I said to him, 'How d'you *know*?' And he said, 'Well, they used to be my trousers.' And I said, 'But they're mass-produced; how d'you know?' He said, 'Look here,' and he showed me a handmade pocket in the trousers. He said his mother had made it from this flowery fabric, because there were no pockets in the trousers and he had to carry his identification documents with him everywhere. And he recognised that unique item of the clothing.

In another instance I had a young woman come around and identify the hand-knitted socks and hand-knitted sweater of her brother . . . [*his voice breaks*] Of course, to see families identify clothing in that way is very sad.

It's terrible . . . I mean, what you're witnessing is the evidence of mass human cruelty – how do you cope?

It's relatively straightforward to deal with the dead in an objective manner, firstly because in medicine we're trained to develop some detachment without losing our compassion. Secondly, of course, forensic pathologists like myself have been doing this work for many years and it's, if you like, *instinctive* to us, it's part of our professional mode of operation. So in a different environment, even though it's a very tragic one with large numbers of people dead, it is possible to operate in that way.

What becomes very difficult – and it's the same for everybody – is when you're faced with the emotion of survivors. That's the same in hospital practice, of course: the most difficult thing is not to deal with a dead patient but to deal with the grieving relatives of that patient. That's always disturbing. So my most vivid memories of these events are always to do with the living, not to do with the dead.

Are you taught how to deal with that side of things when you're trained? Or have you just learned that over the years?

Traditionally we never taught doctors how to deal with these emotional problems. We made the assumption that doctors would acquire it by osmosis and would have strength of character – which is, of course, a nonsensical approach. Nowadays we teach medical students – the young doctors – how to deal with these things, and our education is much improved in that respect. But, of course, I was brought up in the old school of: throw him into the pool and he'll learn to swim! I deal with it reasonably well, and I've coped well over the years, but I still feel the effects sometimes.

Every Christmas, for example, I remember an elderly Palestinian man who was watching the exhumation of the body of his son. It was at night in a small village in Palestine, in the hills, and he said nothing; he just knelt down at the railings round the graveyard, and he clung to them so that no one could remove him. Of course, he was dressed in traditional Arab dress, so whenever I see a nativity scene just before Christmas I remember this old guy . . .

One of the difficulties is that the memories linger with you and you accumulate large numbers of them. A useful way of dealing with them in a professional sense, I find, is to talk about the cases professionally and to add the human anecdote as part of the lecture. By sharing it with large numbers of people, I unburden myself of it.

How have your experiences influenced your work and how you actually do your science?

Well, I suppose you can look at improvements in doing the work from two points of view. The first is the purely scientific. But, of course, we don't do the science for scientific reasons, we do it for social reasons, so the more important aspect is how you improve the social effect of work. That ultimately means how you deal with the community beyond your immediate professional relationships, for example relatives of people who have been killed, victims of violence, people who may have been tortured, victims of miscarriages of justice, families who want information about what happened to a loved one – how that loved one came to die, what evidence there is about the circumstances and from the autopsy that will give them an insight. Sometimes a loved one may have committed suicide, and the family is trying to come to terms with that fact or is denying that fact and wants to explore the issues surrounding it. These days we have a lot more contact with the wider community than we did when I started doing this work. Then it tended to be more isolated, more bureaucratic, and far less compassionate and concerned with the living.

This is very interesting, because some of the older pathologists I've talked to say they went into pathology because they'd had distressing incidents with families of people who had died, and they thought it would be much easier to work behind the scenes. But you're saying that nowadays you're very much part of the social scene around illness and death too.

Well, absolutely. These days forensic pathologists – autopsy pathologists generally – are talking to the families of the deceased. We're talking to a wide range of people, and far from being isolated in the mortuary we seem to spend most of our time talking to people about what we've done in the mortuary! The days where you could do an autopsy, file your report and put it away are long since gone. There'll be telephone calls and someone will ask to meet me. Maybe it's a clinician who's going to phone for the results of the autopsy, perhaps it's the procurator fiscal to say he'd like me to meet with a family. Serving the community in a very direct sense is very much part of what we do.

Why are the general public so deeply unaware of the role of pathologists? Because you've outlined a whole lot of roles – the one that underpins clinical practice, but also the one which gives ordinary people some handle on this whole area of dying, and we'll all experience it in our families at some point. Why are they so unaware of that?

People are aware of forensic pathology on television in an *entertainment* sense, but they're unaware of it in a direct sense because of the way we deal with death in our community. There was a time when most people died at home, their bodies were kept at home until they were buried, and everybody had seen a corpse. These days most people die in hospital: 60–70% of the population. The body doesn't come home; it goes to a funeral director's. So we have less contact with death, and we tend to insulate younger members of the community from death as if it were something to avoid. We tend not to talk openly about death and matters related to it, which, of course, is central to the forensic pathologist's work.

What was your own first experience of death?

It probably would be the burial of a great-uncle whose funeral I went to. I must have been quite young – maybe 11 or 12 years old. I remember the singing at the funeral – it was a Welsh funeral – and that was very lovely indeed. I can still remember standing near the grave, so it must have made a very big impression on me. But my first sight of a dead body wasn't until I went to medical school – I would have been about 18 or 19 years old – and we dissected a body.

In those days we used to have a group of half a dozen people; you'd be allocated a whole body, and you'd spend a year or perhaps longer dissecting the body. It was really a sort of social event: the six of us would be sitting around the body, and as we dissected it we would have all sorts of social and political and academic discussions, on a wide variety of things totally unrelated to anatomy, over the body. You developed quite a nonchalant, in one sense, and intimate in another sense, relationship with a specific corpse. So you become very used to seeing a body.

It's interesting, this, because now people are super-sensitive to the whole idea of the respect shown to bodies. Yet people involved in attending accidents and things talk of the black humour they often use at such times, and medical students tell the same sort of stories as you've just told. Is this how you deal with distressing activities, and does it matter?

My recollection from my university days was that there was never any disrespect towards the body. In fact, the corpse was almost a part of the group, if you like, albeit one that didn't participate in our arguments about the politics and social life of the university and all these academic things that we talked about. But it was very important to us and very precious to us, because we were learning anatomy – we were fascinated by the science and without that body we couldn't learn. Of course, we told jokes, lots of jokes, but they weren't against the body. They were very often totally unrelated to medicine. It wouldn't be uncommon for someone who was walking past to see us standing around a dead body, dissecting and laughing away. Of course, you might misinterpret that as disrespect, but there was no disrespect there. And I'm sure that anyone who had the insight and courage to donate their body for anatomical dissection would be the first to appreciate that.

Respect has no direct relationship with being grim! We show respect in our everyday lives to people, and we can laugh and joke and engage in all the other interactions that are part of being human, and no one would say that was disrespectful. Sometimes glum silence itself is disrespectful socially. So, of course, if you were to see an autopsy today, you would see the pathologist and the technician who were performing the autopsy and at the same time perhaps discussing a football match, or some political problem with the administration, or complaining about the weather, or telling a joke they'd heard in the pub last night, or discussing the soap opera they might be missing because of running late at work – all of these things that are a normal part of human dynamics. If we didn't do that, how much more difficult would it be for us to actually perform this task on a body? It would become unbearable for us.

I suppose in a way it's a bit like an Irish 'wake' – they're anything but grim!

Well exactly; the Irish wake is a very good example of putting a positive slant on something that is very difficult to deal with. Of course, we don't have wakes in mortuaries; I think we travel a middle ground between, on the one hand, not being disregarding of the body, and, on the other hand, not being glum-faced and silent in our examination.

We've talked about your relationship with the relatives of victims of violence, but what about the politics of it? When you attend the scene of a massacre, for instance, how do you cope with your political feelings at the same time as the science?

I've learned over the years that if you want to be an advocate, you must be an advocate, but you can't pretend that you are, at the same time, an objective scientific expert. You have to be one or the other. So I play different roles at different times. I was a founder member and past chairman of Physicians for Human Rights in the UK. There, I was an advocate. But if I go as an expert with a professional group – whether I go to examine bodies or to advise on the results or organisational structure of an investigation – then I'm objective, I'm a scientist.

Advocates have to inject passion into what they do; it's part of the way you must present things. As a scientist you have to leave out a lot of that passion in order to ensure that people look to the facts of what you're saying, and appreciate that your application of the logic to those facts justifies your conclusion and is not clouded by emotion. So leaving out the emotional language is a very important part of being the expert.

When you go to something like a mass grave, do you go with a lot of questions or with a completely open mind?

No one approaches any investigation with a blank mind. Everybody comes with some preconceived notion. The important thing is to recognise what your preconceived notions are and to handle them in an appropriate way and not to be blinded by them. Sometimes the preconceived notions are very reasonable. If you were to go to a grave site in Bosnia and you were to find 40 bodies, your immediate suspicion must be that this was a mass killing as part of a genocide – that would be entirely reasonable. And you would plan on the basis of that, but it may be that you would discover otherwise in the course of your investigation. It may be that these were natural deaths in a hospital, which happened to have been buried in a mass grave because of the circumstances of the war. You would have planned on the basis of your initial impression, and must keep an open mind in relationship to the facts that you discover, changing your plan as new facts come to light.

So have you sometimes found that your initial conjecture was completely confounded when you investigated a body?

Yes, absolutely. I can remember vividly one mistake I made when I looked at a death that appeared to be a suicide. I concurred with the police that all the evidence seemed to point to suicide, and we wrote the case off as a suicide, only to be confronted six months later by the confession of the person who had committed the homicide! So, of course, sometimes you get it wrong – it's just the same as any other aspect of human endeavour.

A final question about the science. How easy is it to be conclusive about a cause of death if a body has been, say, a long time underground?

The more a body deteriorates, of course, the more evidence is destroyed and the more difficult it becomes to determine not only the cause of death but also the circumstances of death, identity and all the other things we're interested in. So bodies that are recovered from mass graves, or bodies that are discovered after many years, present real challenges. Sometimes it's not possible to establish with absolute clarity what happened. Sometimes it's not possible to establish anything at all, not even the identity.

But even the circumstances of discovery of the body can be powerful evidence. After all, if you find 200 bodies in a grave hidden in the woods it speaks rather eloquently of the circumstances by which they came to be there. Very often it's circumstantial evidence in those cases that's more powerful than the scientific evidence. The science is only part of the investigation, it's not *the* investigation. And although we can do some very wonderful things with the science, we're only part of a bigger team with the same goal.

MAD COWS AND HUMAN DISEASE

James Ironside
*Professor of Clinical Neuropathology, University of
Edinburgh*

James Ironside carried out the autopsies on some of the first victims of 'new variant' CJD [Creutzfeldt-Jakob disease] associated with mad cow disease in the mid-1990s and was the first to draw attention to the extraordinary pathology he found in their brains. They had clumps of protein that seemed to 'bloom' like chrysanthemums, he says, '. . . surrounded by a zone of spongiform change – a destructive halo of holes'. Such pathology had never before been seen in the Western world, though it was familiar to pathologists working with victims of kuru, a brain disease associated with ritualistic cannibalism, in Papua New Guinea.

Ironside is a member of the National CJD Surveillance Unit, set up to monitor the fall-out from the epidemic of mad cow disease in the UK, and is today recognised worldwide as an expert in CJD and other 'spongiform encephalopathies' – brain diseases believed to be caused by a rogue version of one of the body's own proteins, called a prion. Huge questions remain about prion diseases: are there still people incubating CJD from eating contaminated beef perhaps 20 years ago? Who else might be at risk of the diseases today, and from what sources? Even 'the prion hypothesis' remains controversial.

I came from a very ordinary working-class Scottish background. No one in the family had ever been to university before. When I was at school my thoughts were initially to do science, because that is what I was good at. I enjoyed chemistry in particular. And then I discovered, in my final year in school, that I actually *hated* organic chemistry – all these chicken wire sorts of diagrams! I couldn't stand it. So I thought, okay, I'll do something else vaguely scientific, and the idea of medicine came up.

Where did you grow up?

I was born in India, in Calcutta. My father was working in the jute industry – he was from Dundee. We came back when I was about five, though I still can remember living there. I remember being on the verandah outside our apartment, and there was a large population of monkeys in the trees that used to come in at any sign of food – they'd just swipe it and run away. I remember going out as a little boy to see an elephant, and having a pet goose, and a drum, which was just a biscuit tin on a string . . . A lot of unrelated snapshots really.

Does it have any bearing on your life now, having had your early life there?

Has India had an influence? I suppose the feeling that you've known something else, that's the main thing, and the awareness that there are lives out there that are very different from what we have here.

I went to secondary school eventually in Dundee, when my parents came home. I discovered that in order to do medicine I had to do biology. At that time at school, for some obscure reasons, you couldn't do physics, chemistry and biology. So I did a crash course in higher biology in a year, and got into medical school in Dundee.

Medicine was really very interesting, but it soon became clear to me that I was more interested in the *mechanisms* of disease than in treating patients. In Dundee you had the opportunity to do an intercalated degree – you could take a year out of the medical course to do something else that interested you – and I chose pathology, because the course sounded specially interesting. You did a bit of pathology, a bit of haematology, a bit of genetics, and I found I was really very comfortable working in that environment. Being a pathologist was almost like being a detective – you're presented with a challenge: here is a biopsy or here's someone who's died; this is what we think, but we don't know; can you make a diagnosis? As a good detective you have to pick up as many clues from the story you're given, you have your list of suspects, and you have to eliminate them one by one, to end up with the truth.

I went back to finish the medical course and enjoyed it, but I knew then that I wasn't going to be a GP or anything like that. So I did my house jobs and applied for a job in pathology. I got a post in Dundee, and we spent a long time in the beginning being trained to perform an autopsy. In those days there were many hospital autopsies, very interesting cases, and we were given as much time as we needed to perfect those techniques. That has stood me in good stead throughout my career.

When did you actually see your first dead body and what was your reaction?

The first time was when we were learning anatomy at medical school and we had to dissect the cadavers. That to me was shocking at the time, because in biology we'd only ever done a dogfish, a worm, beetles and various things, but nothing much bigger than that – and nothing recognisably human.

I remember some people fainting. I didn't faint, but I began to see that the only way I could deal with this was to detach myself from the personal perspective and just focus on the matter in hand.

Of course, in some respects a body is a miracle of nature, and if you approach it from that point of view it's very interesting. I must say, though, that I didn't find the anatomy cadavers particularly pleasant, and when I subsequently saw an autopsy performed on someone who hadn't been embalmed or 'fixed', it was much more real. Everything is so much clearer; you can tell the difference between an artery and a vein, a nerve and a tendon, which we found difficult as students. And the processes – the changes you see in the body – are much more vivid and apparent.

What made you decide to specialise in neuropathology?

Well, that was due to a combination of local people and my own interests. The pathology department of Dundee was a good place to train at the time, because they had some very good diagnostic pathologists. They had one person, Bill Guthrie, who was what we would call a 'morbid anatomist'; he spent most of his time on autopsies, and he taught very well. There was also John Anderson, who had a great interest in neuropathology: he and his research assistant, Beth Hubbard, were working on Alzheimer's disease – looking at what goes on in the brain, and how we can measure the shrinkage of the brain. John Anderson's other interest was at the opposite end of life, with neonatal neuropathology. Brain injury is the commonest cause of death in premature babies, and he was trying to understand how it might be prevented. I worked with all the consultants there, but I was particularly interested in the neuropathology: I just found it fascinating.

That's when I decided that might be the area to train in. But there were no training posts in neuropathology in Scotland at that time, so I got a job in Sheffield, at the Royal Hallamshire Hospital. They had a neuropathologist there, Walter Timperley, who was a great character, a blunt Yorkshire man who called a spade a 'bloody spade' and that sort of thing!

Walter was very good for me because he didn't do intensive supervision – I mean, he was there to support when necessary, but he allowed me to make my own mistakes. Not patient-threatening mistakes, it must be said! He always used to say, 'If you're not sure about something, think again tomorrow.' So although there's always anxiety to get results out quickly, it's actually in everyone's best interest that you don't rush a diagnosis.

In Sheffield I did research mostly on brain tumours, because there was a lot of clinical interest from the neurosurgeons, and we had lots of specimens. We had a multidisciplinary team, and we did some good stuff.

I completed my pathology training in Sheffield, and then a post came up in Leeds, which I applied for and got. But when I started in Leeds there was no professor in post and we had two years without one, which wasn't ideal. I was working with one other neuropathologist, but he then decided he wanted a career change, so for a while I was left on my own. To be on your own as a young consultant is really not good!

It was pretty scary stuff, was it?

It was scary stuff. They had a commitment to muscle disease in Leeds, which I hadn't seen much of, so that was one area where I felt particularly pressed. One occasion I remember, we received a muscle biopsy on a newborn baby, who was floppy and not breathing properly. We looked at this biopsy and I thought: well, it looks like a case of a particular disease called nemaline body myopathy. As it turned out there was a family history of this disease, but a firm diagnosis had never been made. So I did everything I could, including electron microscopy, made the diagnosis, told the clinicians, and I was shocked, *really* shocked, a couple of days afterwards when I asked about the case and was told, 'We switched the ventilator off.' I thought: my God, this is serious. I had discussed the case with other people and all the rest of it, but that made a huge impact on me.

Why had they switched the machine off?

It was a dire diagnosis, a non-recoverable condition. A previous child had died a few days after birth and obviously the clinicians had made a decision with the parents, who felt that history was repeating itself, that rather than go through weeks of prolonged ventilation with no improvement, this was probably for the best. One respects that, but dealing with that sort of situation for the first time was really quite shocking.

So what did that teach you?

I think it taught me that as a pathologist you have always to be aware of the implications of your diagnosis. It's important to remember that diagnosis is not made in a vacuum, or it's not something that exists only between you and the clinician – there's much wider impact.

That matter has been addressed, and most pathologists now work in multidisciplinary teams. Each week here we have a meeting with the doctors dealing with patients with brain diseases, and we review every patient – the clinical features, the brain scans, the biopsy and the diagnosis – and then the treatment is discussed, so we're much more aware of the implications.

What else sticks in the mind from being at the sharp end when you were young and inexperienced?

Well, I remember one patient – and this is where a link with CJD comes in . . . Patients came into hospital the evening before I did the muscle and nerve biopsies, and I used to go and explain to them and the relatives what was happening so everyone was aware. This particular time I saw one poor patient who looked really terribly ill. I suspected he actually had a brain disease rather than just a muscle disease, and in fact he turned out to have CJD and died. I performed the autopsy later, and that was difficult because I had seen him when he was alive and spoken to his wife and other relatives. That was one of my first exposures to CJD, and it was quite dramatic because he deteriorated and died very rapidly, as they sometimes do.

Tell me how it was that CJD became such a central focus in your work?

Well, Leeds was an interesting place to work for a whole range of reasons – not least that we had Richard Lacey. He was professor of microbiology there, and he was a somewhat controversial figure given to stirring things up about listeria and salmonella, etc. BSE [bovine spongiform encephalopathy] had started in 1985/6, and it was interesting and rather worrying that all these cattle were dying. Professor Lacey was keeping a careful watch on the BSE situation, and just before I moved to Leeds, the first case of a similar disease in cats was identified – feline spongiform encephalopathy – and I can remember discussing that with him and his senior registrar, Stephen Dealler. As specialists in infectious diseases their feeling was: well, cattle a few years ago, cats today, humans next year or sometime soon? They had no idea what caused these diseases other than that it was a very unusual infectious agent. Of course, 'the prion hypothesis' was out there – it was 1982 when Stanley Prusiner [who won the 1997 Nobel Prize for Medicine] first proposed that – and it seemed very interesting, but difficult to get your head round.

What is 'the prion hypothesis'?

The prion hypothesis basically states that the infectious agent in CJD and other spongiform encephalopathies (so called because of the spongy appearance of the diseased brain) is a modified form of a protein that occurs naturally in the brain and in other tissues. This modified, or abnormal, protein accumulates in the brain, and this accumulation occurs by progressive conversion of the normal protein into the abnormal form. It's like a domino effect: once you have the first conversion from normal to abnormal, then you get a sort of chain reaction building up as more and more normal protein is converted. The abnormal protein is toxic and damages the brain and eventually causes death. This is biologically heretical because all other types of transmissible diseases rely on genetic material in order to spread, and there seems to be no need for this here.

Anyway, I moved from Leeds to Edinburgh in 1990 and it so happened that that was the year the National CJD Surveillance Project was starting. My colleague, Bob Will, was the physician responsible for the clinical part of the project. We had managed to get funding, and the biggest challenge at that stage was finding space in the hospital where we could work: everyone thought it was genuinely interesting, but they didn't want it anywhere near them because of the perceived risk of 'these prions floating down the corridors and jumping down your throat', or whatever!

After lots of negotiation with the hospital, we were offered accommodation – with very few windows! We started with a small team, just five of us, and at that time it was sold to me as 'a very interesting but very rare' disease that probably wouldn't entail a great deal of extra work.

There had been previous surveillance of CJD, which Bob had been involved in at Oxford when he was a trainee. They'd been particularly interested then in looking at the risk factors for CJD, because the disease, in the mid-1980s, had just emerged in recipients of human growth hormone, and the question was whether any other kind of treatment might predispose patients to the disease.

Aren't there different forms of CJD: 'sporadic', where it happens out of the blue, as it were; and 'familial', where it's inherited?

That's right. There's also the so-called 'iatrogenic CJD' – that is, transmitted accidentally by medical procedures – and the human growth hormone recipients are in that category. Then there are the very rare familial cases such as GSS, or Gerstmann–Sträussler–Scheinker syndrome, which usually affects the cerebellum [the hindbrain, which controls balance and coordination] and looks distinctive down the microscope: instead of spongy change, you have these 'plaques', which are just aggregates, or blobs if you like, composed of the prion protein. The incidence of those cases all lumped together was less than one per million people per year, so I was told there were 50 or 60 cases maximum per year in the UK. But, like everything else, once you start looking you find more.

Probably there was an under-ascertainment in the past, and by the time our project started there was also the 'unknown' element of BSE, or 'mad cow disease' – we really had no idea what that would mean in terms of human disease. So we began by exploring the parameters of what sporadic CJD was, clinically and pathologically. We explored particularly the human growth hormone-associated cases because, although they were acquired not by ingestion but by injection, there still might be some similarities. There was information then emerging on the genetics of the inherited cases too: in the early 1990s there were the first reports of a mutation in the prion gene associated with GSS.

So there was a lot happening, and it was an exciting time. We began to put parameters round these things so that if something different came along we could say, 'Well, this looks different, and it's outside our experience of sporadic CJD.' And indeed that's actually what happened.

Tell me about it.

Well, there were two cases in England; we'd reviewed them and thought they were very strange, and they were reported as letters in *The Lancet* as 'unusual cases of CJD in young patients'. But that wasn't enough to make it a 'new disease'. I suppose what convinced me was the day that I did two autopsies myself. I knew that we'd studied everything, taken everything we could at autopsy to allow us to identify this and gathered all the necessary information. And these two cases: you could line them up side by side and not be able to say which was which, they were so similar. That was the time I really thought we had something new.

It looked in some ways like a combination of GSS, which is the inherited one with the protein plaques, and sporadic CJD, which tends to have a lot of spongy change, because it had both. And the most striking thing to me was using the technique immunohistochemistry to demonstrate the abnormal prion protein in the brain, because you saw a far more extensive accumulation of the protein and a pattern of deposition totally unlike anything we'd ever seen before in a patient with CJD. Over a short period of time we had 12 cases that were published and that all looked very similar. By then there was little doubt that it was something different.

But in order to convince both myself and eventually others, we employed other techniques. One of our strategies was to employ computer-based image analysis. We had a computer scientist working with us in order to do this, so rather than just taking my word for it, they could actually count and measure the amount of abnormal protein and the severity of the spongy change. And on the day in 1996 that Bob Will and I were summonsed to the SEAC (Spongiform Encephalopathy Advisory Committee) meeting, I was able to show them not just a visual depiction of what it looked like, but actually some hard data, and say, 'Here are the measurable differences between these cases and other subtypes of CJD.'

I think that using this hard evidence has enhanced the status of pathology. Fashions change in the science. Until recently it was all molecular genetics: you had to get your bit of DNA, your RNA and then your protein. But now we're in what's called 'the post-genomic era', and we know that actually it's not just having an altered gene that's significant, it's having the gene produce something – a protein that's in the cell and that you can detect. And I think one area that's used this particularly is neuropathology, in looking at degenerative diseases of the brain. All these diseases – Alzheimer's, Parkinson's, Huntington's, motor neuron disease, CJD, the list goes on and on – are associated with accumulations of abnormal proteins in the brain, and we now have tools to study them in far greater detail.

Also, knowing so much more about human diseases, we can be a bit more critical about some of the animal and other models that are used. There was a famous paper published a few years ago with the title, 'Does my mouse have Alzheimer's disease?' and the answer was, 'No!' [*laughs*] Just the fact that it's got one of the genetic changes associated with Alzheimer's and it's churning out a lot of this abnormal protein doesn't make it Alzheimer's disease. We've learned a lot through cell culture and animal models, but the challenge of understanding these diseases in *humans* still remains.

Tell me, when you, as a specialist, see sporadic CJD, does the pathology look the same in most cases, or is there a range of manifestations?

There's quite a range of pathology; even going back to the descriptions of the 1960s it was known there were different *subtypes* of sporadic CJD. There's the type associated with blindness, for example, the so-called Heidenhain's variant of CJD. There's the Brownell–Oppenheimer variant, which is associated initially with unsteadiness rather than dementia. And there's the classical one, which is rapidly progressive dementia. In both clinical and pathological terms those are all different, and what we've been interested to find out is what the basis of those differences is, and clearly it's partly genetic.

We don't know what the cause of sporadic CJD is. One theory is that it arises *de novo* from a spontaneous event in the brain analogous to cancer developing *de novo*. But most cancers develop due to some environmental or genetic trigger, or a combination of factors, so there's a big ongoing study here on sporadic CJD to look at risk factors.

The iatrogenic cases – how different were they from what you were used to seeing in sporadic CJD?

They were different. The patients that developed CJD through growth hormone treatment had a longer illness than sporadic CJD. It tended to present with an ataxia, an unsteadiness, instead of this rapidly progressing dementia. And the pathology of the brain was different. We were interested in the question: how does the infective agent get to the brain? If it's injected into the body, does it go through the blood? Along the nerves? Or does it get there some other way? Our best working idea is that it travels along nerves into the spinal column and then up to the brain.

In the type of research you're involved in, do you use your archive much?

We use it the whole time, but I guess the best example is when we were involved in the identification of variant CJD associated with BSE. As I've described, we had cases of this strange form of prion disease occurring in young adults, which had a particular type of pathology with lots of protein plaques in the brain. So we took great trouble to go back through all the archives – not just our own archives here, but in the literature also – to identify reports of CJD in young people, in the UK and overseas. Talk about detective work – it was really intense! We felt the pressure was on for a whole range of reasons. If you're going to claim that you've found a new disease that's related to BSE, that's no small claim and you want to make sure you've got it as right as you possibly can.

When the BSE problem arose, how much did you already know about other transmissible spongiform encephalopathies – in Papua New Guinea, for instance?

Oh, 'kuru'? Yeah, absolutely, that was very important. Kuru is another of these human brain diseases that's acquired by ingestion, and we had the opportunity to study cases in pathology terms through collaboration with colleagues in the University of Melbourne. Colin Masters, the professor of pathology there, worked with Carleton Gajdusek in the past – Gajdusek was the one who first identified kuru in Papua New Guinea, and received a Nobel Prize for this – and they had some tissues in the University of Melbourne. One of the neuropathologists there, Catriona McLean, came to Edinburgh with kuru material, and we did some comparative studies.

What exactly is kuru?

Kuru was an unusual neurological disease that was a common cause of death in a tribe called the Fore in the highlands of Papua New Guinea. The disease was transmitted within the tribe by the practice of ritualistic endocannibalism, when members of the tribe ate the bodies of the deceased as a sign of respect. The cause of kuru was initially unknown, but when the neuropathological similarities between kuru and scrapie – a common prion disease in sheep – were recognised, kuru was transmitted experimentally and shown to be a 'transmissible spongiform encephalopathy', or prion disease.

I thought kuru would be as different from everything else as variant CJD is, but actually it's more like sporadic CJD than anything else. And that's presumably where it originated from: there must have been a case of sporadic CJD in the tribe and it got passed on because they practised this ritual cannibalism. People have questioned whether it's ethically okay to say that it was due to cannibalism; but having spoken to people who were there at the time, they are in no doubt that that's what happened. So you're not making a moral judgement; it's just a statement of fact.

Kuru is almost extinct now, but what it has taught us is that you can have a long incubation period – of 40 or 50 years – for these diseases following oral exposure. The other thing kuru has taught us is the influence of genetic variability, and that certain types of genetic make-up are associated with long incubation periods. That's why in variant CJD the question is: are the cases we've seen so far simply in the group with the shortest incubation periods? And will other waves emerge later, as they have with kuru? We don't know; we're just waiting to see.

So what are your other big research questions at the moment?

We're interested still, in terms of variant CJD, in how many people there are out there who are infected and that we don't know about. That's still a big question. And what can we do about this? The UK Department of Health has a very large project going on to address this, but they're contemplating another one which, interestingly, is based on collecting tissue from autopsies and trying to see whether any of this tissue, perhaps a bit of spleen or tonsil, is infected by prions. That question will occupy us for a while.

Last time we spoke, you were sitting on the fence over the prion hypothesis. Where are you now, a few years on, with that?

[*Laughs*] I'm still pretty firmly on the fence, I'm afraid. There's no doubt that the prion protein is key to all these diseases, but whether it's the only thing that's important, we don't know. Prusiner himself says there could well be other things, and has postulated the existence of something they call 'protein X'. Sounds like something out of *Star Wars*! But they still don't know what protein X is, and until we have a better idea about what else apart from prions might be involved, I like to keep an open mind.

That's one of the things that drives our interest. We've now started a cell culture facility in our new lab downstairs, and we're collaborating with the Centre for Regenerative Medicine at Edinburgh University, which is very exciting. We're working on stem cells with them. The idea is to develop a cell culture system that we can grow prions in. Once we can do that, we can start looking at what the prions are doing to the cells, and how the cell is reacting to them. Then we can try to stop it; try to block it; try to find out what else is involved. I'm very excited about it.

Tell me about your involvement with the World Health Organization.

I remember particularly going there just after variant CJD was announced. Bob and I were summoned to this big meeting at WHO and basically asked to account for why we thought this was a new disease. All these experts from around the world were there, sitting in judgement on us, and I thought: if we can handle this we can handle just about anything!

It was a slide presentation, and we also had to describe how we'd done the research. I was quite apprehensive beforehand, but, having been grilled in the UK by SEAC and others, I had a forewarning of what might be asked, and could try to offset that by giving the information in advance and showing the hard data from the computer-based analysis. 'This is not just my aesthetic eye saying, "This looks different from that," but there are measurable differences. That is hard evidence.' I think we won them round in the end.

And that put variant CJD on the map, did it?

Yes, it did really; it became a global matter then.

Turning to philosophical matters, tell me, has your long experience of looking death in the face drawn the sting somewhat, or do you fear death as much as the next person?

Dying is an inevitability and it's not something I'm afraid of at all, no. However, the process by which you get there is a concern. Like many people, I'd rather go quickly and cleanly than through a prolonged, agonising illness. And certainly having seen individuals with something like Alzheimer's . . . I mean, CJD is a *terrible* disease, but at least it's relatively short. But Alzheimer's disease is a slow decline, and the burden that places on families, carers, finances . . . I hope that doesn't happen.

On a spiritual level, has your work influenced your beliefs?

No, I wouldn't say so at all. I think one's beliefs are based on other things; other things bring you to ponder the meaning of life and death. Music is one of the main things for me.

For the past few weeks, as part of the Edinburgh Festival, there has been a series of concerts in Greyfriars Church at 6 p.m. – mostly early music concerts, unaccompanied choral music. And difficult as it has been to get myself up to Greyfriars for six o'clock after a working day – I've been coming in early and doing what I can – to stand queuing in the churchyard near all these graves, and then to get into the church, sit down and just listen, with this wonderful acoustic, has meant more to me than seeing any number of dead bodies. Just to listen to the sound of these voices in this music, and to think: this is the sound of *life*. That's been quite wonderful.

FROM ONE GENERATION TO THE NEXT

Helen Wainwright
Department of Pathology, University of Cape Town, South Africa

When she began studying medicine, Helen Wainwright wanted to be a paediatrician. But during six months as a senior house officer at Red Cross Children's Hospital in Cape Town she caught all manner of diseases from her small patients, and says, 'By the end of the six months I thought: well, I've never actually been so sick – this is not for me!' She turned instead to pathology and has found the exploration of disease processes a satisfying and richly fascinating field. 'As a clinician sometimes you're feeling in the dark: you give a particular treatment and you hope for the best. But with pathology you find out what actually happened – in most cases.'

Besides her duties to teach new generations of pathologists, Wainwright is in charge of fetal and neonatal pathology services for all the maternity units in the Cape Town area. Since the demise of apartheid, the city's population has swelled with migrants from the countryside looking for work. Diseases of poverty are rife, health services are overloaded and the rate of stillbirths is exceptionally high. In the autopsies she carries out, Wainwright finds very many babies damaged by alcohol abuse, diabetes and infections in the mother.

In specialising in pathology, she has followed in the footsteps of her father, who became professor of anatomical pathology at Durban Medical School after emigrating to South Africa from the UK with his family. But Wainwright was unable to train in Durban because the medical school was reserved for non-whites. She remembers how the rules of apartheid affected the practice of medicine too when she began her career.

———

[When I first started in medicine] I worked at Groote Schuur Hospital, and we had a white side and a black side, and they would have different food and things. But your team would have both white and black patients.

Then when I went to Red Cross Children's Hospital, again you had wards which were for the different races. In the middle of the night you got a sick baby, and the one thing you really couldn't care less about on a busy shift was if the baby was white or black – if it's a sick child it needs care, it needs a bed. And then in the morning we had to sort them out: the doctors would change the colour of the folder and whizz the child across to the right section – it was absolutely daft!

But care on both sides was absolutely equal at Red Cross. And the children had the same sweeties, the same little pictures . . .

You've worked under both regimes – what have been the big changes you've seen?

The sad thing is that Groote Schuur used to be a top place. It catered for people who had no money at all, but it was *the* top place for all the wealthy people in Cape Town. The best surgeons were sitting there. So the [rich patients] would come and pay a huge amount of money, which funded all the other activities.

You never had people turned away. Whoever was sick got in, no question. But people who could afford to paid handsomely; so your equipment was very good and you had top-quality care, whether you were black, white or whatever. And the hospital was always spotless – both sides, black and white.

Then Nelson Mandela came in and said, 'No charge for all children.' Now, there were plenty of children whose parents could afford to pay, so that was a shame. He made the statement, and all the people who could afford to pay suddenly stopped paying. So that wasn't such a clever thing to do.

My sister [a paediatrician working at Baragwanath Hospital] in Johannesburg used to get very angry because you got more [of the government health budget] for a white patient than for a black patient in the old days, but if you just shared it between the two, who was to know what you were up to? At Baragwanath, you see, they were all black patients so they had to manage on much lower budgets because nobody was paying.

I think Baragwanath is now getting more of the budget, but they still struggle. If a patient comes in sick, they admit them whether or not they've got a bed. So initially, in the paediatric wings, it was one person in a bed, then two in a bed, then one under the bed, then . . .

But what about here at Groote Schuur?

They've just cut the beds, so now you die in the street.

So you don't admit people here if you don't have a bed?

206

That's right. It's terrible. Absolutely terrible.

Tell me about your day-to-day work here. What do you find when you come in, in the morning?

I go for a swim first, at five o'clock, and then I come in just after 6 a.m., because that's a good time to get work done. Once the day starts at 8 a.m. the telephone rings, you get distracted and you can't do things easily.

Part of our job is to train new anatomical pathologists. The sad thing is that a lot of them tend to push off overseas, so we train them for other countries. But that's their choice. You can't say, 'I'm only going to train you if you stay in South Africa'; that's not reasonable.

Why do they leave? Are the conditions for pathologists here not good?

The salaries in the private sector are very high, and the salaries in the state sector I don't think are bad. But if you're a young person and you want to have a family, then buying a house is difficult . . .

What are your responsibilities besides the teaching?

I'm in charge of the fetal and neonatal service and placentas, so all sorts of interesting things crop up. Then we also have a surgical roster.

How many autopsies do you have per day?

Oh, smallish numbers. A couple of years ago I was doing about 300 autopsies a year. Then we were taken over by the NHLS [National Health Laboratory Service] instead of the Province, and they charge hospitals a lot for the babies, so they've cut back on autopsies. There's a dissection fee of

1,500 rand [about £120] for a baby, but there's no dissection fee for a placenta, so now I've cheated – I have my sneaky ways of handling things – I call the little babies 'placentas'.

We have so many babies born dead that we try and look at the placenta of every single one in our area, which covers all the maternal obstetric units in the Cape Town area.

Why are so many babies born dead? What's happening?

We have huge problems with alcohol, and we have huge problems with young women who don't even appear for antenatal clinic. You don't know whether it's because the antenatal clinic is busy when they try to get time off to go – these clinics tend to take a set number of people, and if you're not there in time they won't look at you. There are all sorts of problems, so a lot of them don't have any care at all.

And then we have a lot of pre-eclamptic toxaemia – a much higher incidence than elsewhere. No idea why. We have a high mortality from it, mothers and babies. Very high blood pressure can suddenly appear, and if you aren't monitoring a person, they don't notice that their feet are swollen and they wait till they start fitting.

That's one of the big things, is it?

That's one of the big issues, yes. And you know, we have this huge fetal alcohol problem, where the mother drinks, all her friends drink, the relatives drink, and it just continues like that. Those babies are all growth-retarded. They don't catch up, and it's a major problem. A certain number die *in utero* or are miscarried early on. But the main problem is the children who have mental retardation for the rest of their lives.

What do you find with fetal alcohol syndrome?

The bad ones you can recognise just by looking at their faces. They have a long, smooth upper lip, a thin lip. They tend to have upturned noses, often a squint, eyes quite far apart – you can take just one look, even when it's a little fetus, and have a very strong suspicion that it's fetal alcohol syndrome.

But that's the tip of the iceberg, the ones you can see just by looking. There's a *huge* number who don't have the facial features but they have the brain damage; they don't grow; they have multiple organ damage and that sort of thing. They're the ones at school with a short attention span, and the teacher doesn't know what to do with them: they're untrainable.

And is the damage you see dependent on how bad the abuse has been at different points of development?

That's right. So with binge drinkers, you're suddenly providing a huge amount of alcohol. But you've also got to take into account how the liver metabolises the alcohol. Some people are very wonderful metabolisers, break it down quickly and get rid of it, and other people are slow, so the breakdown products hang around for longer and cause a lot of damage. That's why you might find people with similar drinking habits and one having severe effects and the other less severe effects.

What are the roots of the alcohol problem here?

People being paid in alcohol, that's the root. When people worked on the wine farms [in former times] part of their wages was alcohol. And it's an addiction – whole communities are caught up in this. And then, if you think of South Africa and Australia, we're very keen on sport, we drink a lot . . . It would be very nice if, when the wife's pregnant, the husband also abstains for the nine months to make it easier for her.

How did you feel when you first started to see all this? Was it quite shocking?

Dreadful. It was shocking. If the brain's damaged there's nothing you can do. This is the commonest cause of mental retardation, and it's preventable.

There are people who feel it's okay to have the odd glass of wine when they're pregnant. But animal studies show that it's a poison, and the brain cells get killed off. You might say, 'Well, we've got masses of brain cells,' but do you want to give your child a decent chance or is it okay if a few brain cells are knocked off?

What other things are you seeing in the babies that you're autopsying?

Well, we're still getting lots of syphilis, and that's absolutely treatable. We shouldn't be getting any syphilis. You go to antenatal clinic and the first thing they test for is syphilis. If you've got it, they treat you straight away with penicillin, and it can be cured. But if you don't treat it, the babies end up with bone deformities and all sorts of horrible things – diseased liver, diseased multiple organs . . .

I think the trouble is that we're having large numbers of people moving into Cape Town, and there aren't the facilities available. It's not like in the UK, where people will only move somewhere if there's a job. Here people just appear, and there aren't enough medical facilities for them.

There's supposed to be free treatment for women and children, and yet not so long ago a mother came to a clinic in the Western Cape, and she didn't have an ID [identity document] on her so she was refused treatment. So she borrowed her cousin's ID and then died as her cousin. No one has any right to refuse a pregnant woman medical treatment because she hasn't got an identity document. But the system is overloaded, you see. And these women working at the maternity clinic say to themselves, 'Well, I'm going to see so many pregnant women,' and then they shut the doors, which is very wrong. But they say, 'Well, I've got to go home and cook supper for my family, and if I try to see too many people I can't cope.' So that's the problem. This lady did eventually get seen, but unfortunately as somebody else. And she died as somebody else. It was very sad.

Last time we spoke, you were quite angry about the state of medical education here . . .

And the medical students we're turning out? Yes, it's very disappointing. When I did my undergraduate training, I think it was very good – equivalent, at that stage, to most other places in the world. Now all the science has gone out of it – pathology forms something like 2% of the total marks in the year. Medical training today is all about being in contact with your feelings. So the next generation of doctors will be very nice and hold your hand as you die, but they won't be doing anything to help you because they won't know why you're dying!

But we have got some very nice pathology registrars, and it's very enjoyable watching them develop. That's an absolute pleasure.

Our registrars have to do 50 autopsies each before they can sit their final exam, and there are very few being requested by the hospital – Groote Schuur has an autopsy rate of about 0.05%. We are trying to make up for this by offering our services to forensic pathology. If anyone is stabbed, or has a motor vehicle accident, or suffers an unnatural death, they go to the forensic pathologists. But they also get all the natural deaths of people who haven't seen a doctor – anybody who just drops dead – and they're very happy to open their doors to anatomical pathologists who want 'natural' deaths. So that's how we're trying to compensate.

And the registrars are really 'enjoying' it, if you know what I mean, because they're seeing pathology that no one has modified by medicine. You see, when we get a case from the Groote Schuur Hospital, the patient might have been in ICU [intensive care] for a month, and by then their tissues are in a bad state. What happened initially is often masked by all the drugs and the ventilation and everything that's happened while they've been in hospital, whereas these other people are just wonderful pathology.

So you literally start with a clean slate? You don't know what this person dropped dead from?

That's right. Sometimes they're in their office, clutch their chest, or say, 'I've got a headache' and then die. Or they're waiting at the bus stop, and they collapse. You might have a little clue, but often you've got no clue at all. You've got to realise that somebody who may look in wonderful shape on the outside may not be in such great shape on the inside! So it's very good experience for the registrars.

But they don't test the bodies for HIV at the police department, so you've got to be very careful doing pathology there. And you've got to be quite a tough registrar to work there, as you see very unpleasant sights. Even though you're working on a natural death, you're in a huge mortuary with large numbers of other deaths, many of which are very distressing to see. And the facilities are grim. The Dark

Ages, I think you'd call it. But if you looked at our hospital mortuary, you might also find that a bit of a shock. Compared to the UK, we're well behind.

We've had some terrible cases . . . A newly diagnosed diabetic woman of 45; she came along to hospital, they kept her in and treated her overnight, and then they discharged her and she died a day later. Twenty years ago we'd have kept her in until everything was sorted out, but now they want the beds for HIV and TB cases. Those are the two things they're good at treating – they're no longer very good at treating these other conditions. So if you've got diabetes – and I'm sure this was a productive woman of 45 – you die, and that makes me very angry. Why should you have to have HIV or TB before you can get decent treatment?

Another one was a 79-year-old man who had lots of diseases, and he was having bowel symptoms. In the notes they said he'd had abdominal pain for two days, and then he died. He'd been given a note to come back to the clinic because they were worried that he'd got cancer of the colon. When our registrar went down to do the autopsy, he found that the man had this huge hernia in his groin, and the bowel had become stuck in there and had perforated. I mean, how could you *not* notice it?

So your autopsies are showing up where there's been negligence?

Well, that's the value of an autopsy. And at a place like Groote Schuur, if they're stopping them, then it means you're not picking up any of the mistakes that are being made. It's one thing to make mistakes when you're working in a little place and you've got nobody you can ask for advice. But we're not even checking at Groote Schuur to see that the standards are being maintained. That's bad.

What do families say when they hear about these things?

I think we've got very forgiving families here, very forgiving.

Do you see a lot of very strange things in the course of your daily work?

Yes, because we're seeing babies who die; we're not seeing the babies who live and where everything's fine. So we look at a lot of stillbirths. We do a placenta on every stillbirth, and if the baby looks peculiar, they try to get permission for an autopsy.

The other day we saw a term baby, and it had a head about that size [*holds up hands to show how tiny*], a single eye, no nose, tiny little mouth – and then extraordinary things going on in the brain. Basically virtually no brain at all. A seventeen-year-old mother; the baby died within a few gasps, tiny little lungs.

Have you seen this before?

I've had about three infants with 'cyclops' anomaly. You don't get a lot of it. Sometimes it's two eyes fused together, sometimes it's just one. One of the causes is chromosomal. And the other is diabetes, because you get all sorts of things with diabetes.

With the new students this week we had an abnormality called a 'mermaid'. It was a baby that looked fine at the head and the arms and then tapered into just a single fused leg with one foot on the end. No genitalia. Oesophagus had stopped up here. [*she touches her chest*] No stomach. Little bit of bowel. No kidneys.

But it was normal from the chest upwards?

Externally, but it wasn't normal when you started dissecting it. With that one, the mother was definitely diabetic. We have a lot of overweight females and a lot of maternal diabetes.

This is becoming a big problem in Europe, but I don't think people are aware of what diabetes can do in pregnancy.

Diabetics in Europe will go along to their obstetrician before they decide to fall pregnant, and the obstetrician will tell them that if they're beautifully controlled when they fall pregnant there won't be a problem. Our people don't plan pregnancies. You're suddenly pregnant and your diabetes is all haywire, and then you get all these malformations. So that's the trouble.

And do you see a lot of these?

Plenty, yes. A lot of them involve the brains and particularly the lower part of the spine. In the last year or so I've had four really severe cases. So it's very interesting.

You're obviously excited by the science of it. But as a mother yourself, how do you psych yourself up to see only the science and not to think of the implications for the mother?

Well, it's interesting. When I see a 'cyclops' it's not distressing to me at all, it's fascinating to think of the pathology that will be sitting in that eye. I mean, there was no lens there, and all sorts of extraordinary things happening.

One of the lectures I'm giving to the students is on abnormalities of the brain, and this one is real, and it's very, very severe. It's the most severe end of the spectrum that I've ever seen.

What are you most interested in? The malformation of the eye?

And the brain. I'm tying them all together and asking, 'Was it one "hit" – did all the problems occur at the same time, or not?' Then, 'At three weeks post-conception, are all these things developing at that time and being damaged? Or did something happen at three weeks . . . Something happen at six weeks?' While all those organs are forming it's a very vulnerable time in pregnancy.

The arms develop before the legs . . . There's a sequence in which things happen, so you can often time the [event that caused the disruption to development] by the pattern you get. Very interesting.

Do you ever see the reaction of the parents to some of your reports? Because I think that if I delivered a baby with the kind of malformations you've described, it would be very difficult to deal with.

No, I don't think it would. I think you'd go on the Internet and you'd start reading about it. My second child, one of my three children, had a congenital heart defect, a squint and club feet.

And how did you feel about that?

Well, obviously one was distressed. I mean, you don't know whether they are going to pull through or not, because some of these congenital things are lethal. With my son I presume it was some sort of infection [during my pregnancy]. There wasn't ultrasound in my day, so there was no question of detecting it early. But these things happen. You just accept it.

My son is 26 now. He's playing cricket. Very awkward, very difficult . . . I'm having a big battle with him now, but . . . !

So has your own experience made you fatalistic, or do you think you always were fatalistic – or philosophical – about the hand life deals you?

I think I always have been philosophical, yes.

Tell me, after you've done an autopsy, how much work is there still to do? With the baby with 'cyclops' anomaly, for instance, what will you do next?

Well, I'll check to see, is it a [chromosomal abnormality called] trisomy 13? I'm going to count up how many features would support [that diagnosis]. They usually have an extra finger; this one didn't have the extra finger. So it's a matter of counting up how many things I've got *for* it [being a trisomy 13] and how many *against*.

What else would you have looked for apart from the extra finger?

Congenital heart abnormality. This one had an unusual abnormality of the vessels coming out of the heart, as opposed to inside the heart. So that is much less clear cut.

What other things are you seeing a lot of, besides complications of diabetes and alcohol?

Well, we still have a lot of rheumatic fever in Cape Town, and if you don't pick it up the person gets valve damage, and often has to have a valve prosthesis. They then have to go on blood thinners, and the easiest one in South Africa is warfarin. If they have warfarin while they're pregnant, then the baby is damaged. It can cause bleeding in the baby, or it can cause malformation. A tiny, squashed little nose is a typical appearance and they have great difficulty breathing. Then on an X-ray they have lovely speckled calcification next to the spine and all the growth plates. The last paediatric pathology conference I went to I took my experience with warfarin, and people said, 'Look! We don't see this sort of thing.'

Is that because in the West they know the effect of warfarin and this would be picked up in antenatal care?

Well, they've tried putting people on different things here, and overall warfarin is best for the mother. You put her on other things that are better for the baby, but the mother dies. No, you want to get rid of rheumatic fever, which is caused by a streptococcal sore throat. If you don't treat that with antibiotics, then two weeks later there are antibodies that attack your heart, your muscles and your connective tissue throughout your body.

But you don't want a doctor who gives antibiotics for every single sore throat he sees, because if you give penicillin for a viral sore throat all you're doing is building up antibiotic resistance in your community. So you want a discerning doctor. Will our new ones be discerning? We hope! But we've brought in medical practitioners to teach in our curriculum, and guess what? They're saying to the students, 'You don't need to know about rheumatic fever because it doesn't exist any longer.' Why? Because they're missing it!

You see, rheumatic fever is very sneaky. It's pain in the joint, and then it can flip to another joint. So you might say, 'Well, this child is just being damn difficult.' But if you listen to the heart, you realise what's going on. If you're not thinking of it you can miss it, and then it quiesces; so if they survive that initial part, then you get all the secondary changes.

So how big a problem is it – how often do you see it?

Well, the cardiac people see it when they get patients with valvular disease. We see it from surgical specimens – the surgeons put in a new valve, they take out the old valve and send it to us. And then we see it in autopsies. So when somebody teaches that it no longer exists, you get a little concerned.

If you let a child who's got rheumatic fever play sport, they drop down dead while they're playing. The other time [people who've had rheumatic fever] get caught is in pregnancy: your cardiac output has to double, so if you have a narrow valve, suddenly your heart can't cope with that extra circulation. We get a few coming in as maternal deaths.

I say to the medical students, 'Well, I'm afraid, you *do* need to know about rheumatic fever, because we see it.'

You've got the biggest HIV epidemic in the world here in South Africa – what are you seeing of that in your work?

With the HIV epidemic we're seeing all sorts of interesting infections – and different ones from those in the UK, because they've now got large numbers of patients on antiretrovirals, whereas we've got large numbers of patients who *aren't* on them.

So what infections are you seeing?

Tuberculosis and tuberculosis and tuberculosis. In every conceivable form. And then sometimes two or three simultaneous infectious diseases – so tuberculosis with various viral things. All sorts of brain pathology. Cryptococcal meningitis is the commonest meningitis now in Cape Town, whereas it used to be very, very rare. Toxoplasmosis we used to see very rarely, if you had a pet cat or something like that, whereas now we see it more frequently. But unfortunately, tuberculosis in *everybody*. We get biopsies from large numbers of HIV-positive patients while they're alive, but unfortunately we do relatively few HIV autopsies.

What are the most difficult moments for you in your work? Irene Scheimberg says she can't bear it if a baby is brought in for autopsy in its clothes. Some things just get through the armour.

Right. I don't like doing an autopsy when you find that somebody made a terrible mistake. I had a baby the other day – the mother I'm sure wasn't particularly wealthy but she was going private, and you expect people to put themselves out if somebody is paying privately for a pregnancy. She said – and it was subsequently reported in the newspaper – that she went to the hospital in labour, the doctor wasn't there and she delivered on the bed with a sister who wasn't particularly helpful. Then the sister took the baby away and didn't tell them anything. I don't know if that is true or not, but anyway the baby was taken off to ICU and the paediatrician wasn't called. I think the paediatrician came four hours after the birth to find the baby in a very bad way, and it died.

I don't have any problem with anyone who's doing the best they can. Medicine is a tricky business and accidents do happen. But when it's just negligence I find it very upsetting.

Why did the baby die?

Well, the obstetrician who asked me to do the autopsy said that the cord was around the neck. There was no evidence of that, and I just wondered if it didn't get the right treatment. That's what I don't like – it was a beautiful-looking baby, and it seemed like an unnecessary death.

Are you a religious person?

No, not at all.

Do you think religious belief is incompatible with what you have seen as a pathologist?

Well, I think for some people a spiritual belief is what they need. I don't have any objection to that. I have one religious son, and to me that's his choice. He's the one with the cardiac problem – he had cardiac surgery when he was two because he had a hole between his two ventricles, and the vessel going to the lungs was too narrow, so it had to be widened. He was in heart failure from about three weeks old.

Why did they leave it so long to do the surgery?

Because the heart's walnut-sized at birth, so they won't do it until it's a reasonable size to operate on. And they did a wonderful job. Once they'd fixed it, he just shot up like a bean! Before, people used to stop me and say, 'Excuse me, why don't you feed this child?' because I had a daughter who was quite large, and this child with chicken-drumstick legs – these two skinny little things!

Those two years before he had surgery, what was life like?

Well, he wasn't allowed to cry. The moment he started crying he just became too tired to drink. He was having difficulty breathing, drinking, anything like that. So we had to cradle him a lot, or put him in the back of the Kombi [van] . . . We had some dirt roads nearby and they were absolutely wonderful – the car would go bounce, bounce, bounce, and he would drop off to sleep straight away. Or you'd put wrinkles in the carpet and put him in the pram – he loved rough things – and he'd drop off. If you sang lullabies it didn't work at all.

So how did you cope? You carried on working . . .

Oh no, I didn't. No, no, I stopped for six years. You know, I enjoyed having the children, and I intended giving up work for an initial period. Those first two years are great fun.

How old was your little boy when you went back to work?

He was four, and he'd had his surgery and all that. His surgery was absolutely phenomenal. They went in through the atrium, the thin-walled chamber, and put a patch over the hole.

And what did they do with the plumbing that wasn't quite right?

Just sort of chiselled out a bit of the muscle to make it wider. It was just incredible. It was done at a hospital called J.G. Strijdom, and the change, you know! He had to go back three weeks after his surgery, and I put him in the car – he was only two and a half, but he could see we were going towards the hospital, and suddenly from his little car seat in the back there was this horrible howling. This little thing with his little cap on – I almost had to drag him up the stairs because he *did not want to go*!

He'd been tied down – they put him on a ventilator, and they tie you down, you know? He was actually in ICU for a relatively short time, but he obviously knew all about it. Then the doctor said, 'I just need a little listen to your heart,' and he listened to his heart and then said, 'Alistair, I don't want to see you . . .' And this child just flew down the stairs, dashed to the car!

The surgeon was absolutely phenomenal. And my sister was there to do the regular check-ups. We'd started saving for the operation from the moment this child was born. Then it was done at Strijdom, which was a government hospital, and it cost next to nothing. So we gave a big donation to research.

But how traumatic was it for you?

It was extremely traumatic. When Alistair went into ICU we made sure he had somebody with him every single minute. You weren't allowed inside the ICU, but you could look through the window and he could see someone of his family there all the time.

But people came up with all sorts of lovely little things [to keep him happy] . . . You didn't need to do much. Then he came home, and within four days of having open cardiac surgery he was on one of those little plastic scooters tearing around the house! [Nowadays] my child with the heart problem goes rock-climbing! I don't want to watch. To me, that's his choice, but to watch it is the worst thing.

Going back to your work, did you, as a woman and mother, ever feel there was a glass ceiling? I mean, what was it like being a woman in pathology?

There are lots of women in pathology. It's good for women. You get the occasional call at night, but nothing like the other disciplines. It's never been a problem for me.

Are you as excited now by pathology as you ever were?

Oh yes. I look forward to coming in to work in the morning, yes.

OF ROUGH POLITICS AND POWERFUL PATHOLOGY

Kumarasen Cooper
*Vice-Chair and Professor of Pathology, and Director of
Anatomic Pathology, University of Vermont*

Growing up in apartheid South Africa, Kum Cooper had to fight enormous odds – including the trial and jailing of his eldest brother for political activism – to get his medical degree. But after a three-year fellowship at Oxford University, where he got his DPhil, he returned to his homeland to become one of the first professors of colour in the medical college at Witwatersrand University in Johannesburg. These were extraordinary times: apartheid had collapsed and all South African institutions were transforming, but race remained a potent issue. 'You do not just flick the switch and say, "Okay, yesterday was apartheid, today we have no prejudices." So I was going to be dealing with people who clearly had prejudices,' he says.

At 'Wits', and later in the US, Cooper set up and ran experimental pathology labs where research questions raised in the day-to-day care of patients could be explored by pathologists in his department. His own special research interest throughout his career has been cervical cancer associated with the sexually transmitted human papillomavirus, HPV, which was rife in South Africa when he began work on the wards as a newly qualified young doctor.

Cooper's personal experience of pathology in the developed and the developing world over more than two decades – as well as his role as a leader of change, particularly in South Africa – gives special weight to his concerns about the state of pathology today.

———•———

I was born in Durban, on the east coast of South Africa. I was born into the apartheid regime and knew no different. You were taught to think of yourself as an inferior being and were treated as such. You went to separate schools, ate in separate restaurants, used separate buses and lived in separate areas. And, of course, you had no political representation.

My father was a farmer and my mother was a teacher. To survive in such a society you needed to be educated, and my parents decided that in order to give their children a proper education they should build a school in the area. The Cooper Government Aided Indian School was partly funded by the government, but my mum and dad got funds initially from the local community, as well as putting a lot of their own money and time into building it. In South Africa at the time there were four different curricula: the African, the Indian, the so-called coloured, and then the white system, so we had to follow the restricted curriculum.

My brother, who is five years older than myself, became increasingly aware of the system around him, and at university he became highly politicised. He formed the group known as the Black People's Convention, or BPC, along with Steve Biko. I grew up in the shadow of his activities, literally, being protected by my mother to ensure that I continued with my education, got into university, and then into medical school.

In those days there were six or seven medical schools in the country, only one of which was for non-white students:

Natal Medical School. To apply to the others you had to get ministerial consent, and I was refused throughout my university days.

Where did your desire to do medicine come from?

I went into medicine primarily because of my mother. I thought I was going to be a biology teacher, but after I went to university she made me believe I could do medicine, so I used to send in these applications really for her, and they kept coming back negative. Then in the fourth year she said, 'Just do me one favour: send in for the Natal Medical School.' And sure enough it came through.

In parallel, my brother's activities had attracted attention from the powers of the day. Around 1974 the Black People's Convention organised a rally to celebrate Mozambique's independence. A whole group of them were arrested, including my two brothers. That was the beginning of the so-called SASO/BPC trials – SASO being the South African Students' Organisation. The trial was known as 'Saths Cooper and Others'. Ultimately, I think, 12 of them were convicted. They served two to three years in confinement during the trial and then a further six years on Robben Island. In fact, my brother's cell was right next to Nelson Mandela's.

My goodness! And were you able to visit him?

Yes, I did, and those were emotionally exhausting experiences . . . being treated in a way that was clearly meant to make you feel inferior. Just flying into Cape Town, getting an appointment, ensuring they approved your visit, then getting on to those boats across to Robben Island, where you were put below deck . . . rusty and dirty; it was horrible.

Those were difficult days . . . Oh yes, I grew up believing the white population was superior to me. My brother didn't share those feelings. The BPC were the architects of so-called 'conscientisation', which literally encourages you to recognise that you're a human being like any other. It was conscientisation that made us all, the youth of the day, begin to realise, 'I'm no different.' When I completed medical school, started training as a pathologist and began to mix with my colleagues who were mostly white, I began to feel part of this human family for the first time and no different from anyone else.

And did you find that low self-esteem affected your training?

No, the educational task was clear: I wanted to do medicine and I was focused, so that was not a problem. But, having said that, there were outside interferences, because this was a period of turmoil in South Africa: this was the 1970s, and you may recall the Soweto Riots. Our only means of demonstrating our disapproval was to boycott our studies, and that severely restricted my education. From 1972, when I first went to university, I think we spent about 25–30% of our time marching and demonstrating.

And how did you feel inside? I mean, with your passion to do medicine and also seeing what was happening in the country?

[*Reflective*] Nobody has ever asked me that question . . . It was an extremely difficult period. My brother was a leader of this movement. My quest to educate myself – to realise my mother's vision – brought a lot of internal turmoil. You'd be trying to concentrate on your work, but during the day you'd be sitting listening to protest songs and not attending your lectures as you should. Throughout my university and medical school days, that literally peppered my education

– right up until the final year, when we boycotted more than 30% of the time. We stood extremely close to losing the entire year, and it was by sheer act of will that we managed to write our final exams and get through. It was *extremely* difficult.

We didn't talk to each other at home about these things because of the conflict, you know? Mother wanted you to get on with your studies and graduate. You were on the verge of becoming a physician . . . You can imagine the anxieties for a parent. But I was always able to study, thankfully, and I think that probably carried me through many of the turmoils of my life.

And there were significant upheavals: when I was in the final year of high school, my brother was already at university, and a few days before my final exam, my matric, he was arrested. The security police came stomping on the door at three in the morning. These were huge, monstrous guys. And they were rough. My brother has never shared with me the brutal times he spent in solitary confinement.

So you didn't have big political discussions at home?

Not those that you'd expect a family like ours would have. It was strange. There was a lot of conflict because my mother would have her say about my brother's activities, you know: 'Why did you have to do that . . . inviting all of these problems?' Those are the family moments that I recall, and a lot of turmoil. But the solid foundation of the family was always my mother.

You say you thought you would become a biology teacher – when did you start to show an aptitude for science?

In high school. I just loved the laboratories, the science and the experimentation. Those masters took us to heights we couldn't believe we could achieve. But the order of the day was that no non-white student could ever get beyond a 'C' grade, so 'C' was a merit – it was highly acclaimed. I was no way a 'C' student, I should say! But 'D' got me to university, and my grades eventually got me into medical school.

And I loved medical school – the knowledge and understanding of disease processes, translating that into clinical signs and symptoms, and seeing patients; it was just fantastic. Pathology was taught over two years, and we had the luxury of having English teachers. Professor Braithwaite was my professor of anatomy, a marvellous gentleman! He was an artist, literally, drawing these anatomical structures on the board that were almost photographic – that's what you took in and kept with you when you dissected a body.

I loved every minute of learning anatomy. And Professor Wainwright, from Sheffield, was my first professor of pathology. He was really a marvellous soul. Pathology was *extremely* well taught. We attended three autopsies a week – full autopsy, from the moment of dissection to drawing up a clinico-pathological profile of the entire case. It was superb. And we had practicals where we looked at slides and bottled specimens. I think we were spoilt. When I think back now, nowhere compares to the training I and the people of my time received.

As I said, I'd had the idea of teaching biology, so the ability to marry the love of teaching with the love of biology attracted me to pathology. You saw patients as well, and then later you had the opportunity to interact with all the physicians in a variety of disciplines, whether it be the dermatologists with a skin biopsy, or an orthopaedist who was repairing a bone in the joint.

So you were not just backroom boys?

Oh, no. We had regular clinico-pathological conferences and discussed individual cases. The hospital was just yards from the pathology department. These days I think the department is more isolated, and surgical pathology is practised behind 'the paraffin curtain', as it were, with the clinicians the other side. The training was multidisciplinary. We did – and still do in South Africa – train surgical pathologists to practise the art form in all disciplines. It's not like in the United States, where people would then go into gastrointestinal or brain pathology, or head and neck pathology or soft tissue pathology. In fact, this is a concern of mine – that the classical surgical pathologist is a dying breed.

When you were working as a pathologist in Durban, what was your caseload?

We saw everything. You can talk to pathologists in developing countries and listen to the wide array of fantastic pathology – not 'fantastic' from a patient point of view, but from a learning/training point of view. The range of it, the depth of it, was just surreal. For example, I spent a week in Durban this summer and I was able to get about 12 to 15 'cases' (forgive the term) of pathology on slides, which are easy to transport, and which I could bring back to Vermont to teach my residents. Now those 12 to 15 cases of fantastic pathology I would probably take about a year to see in Vermont, but I saw them in a week in Durban. That just gives an idea of what we were seeing: unusual tumours; diseases you rarely see in First World countries.

Anyway, I completed my five years' training in pathology, and then I did another three years as a consultant. And during that period I applied for, and received, a Nuffield Fellowship to study in Oxford. It was very competitive and prestigious, and I was fortunate to receive the Fellowship in 1990.

Had you an idea of what you wanted to study at Oxford?

I had. One of my second choices for a discipline, apart from pathology, was obstetrics and gynaecology. I just loved the dynamics of obstetrics and the labour ward, and I was fascinated by the gynaecological oncology. During my internship, which was two years, I'd had a lot of patient exposure. There were five gynaecological wards, and every time I'd go into one I'd find 20 to 25 patients with cervical cancer, ranging in age from their twenties to sixties, who had come in for staging and grading, and then they'd go for radiation or surgery. It was during that time that the notion of the human papillomavirus as a causative agent for cervical cancer was being suggested.

Cervical cancer was known to be a sexually transmitted disease a long time ago. There was the notion of the 'male factor' – the scenario of the wife who stayed home and remained faithful to her one partner, but whose husband was not faithful, and then she developed cervical cancer. So it was known to be sexually transmitted. But as to the agent itself, suggestions varied from cytomegalovirus to herpes and all kinds of organisms, until a scientist in Germany called Harald zur Hausen determined that it was human papillomavirus. This was the early 1980s. I'd learned about it in medical school, and then as a pathologist I'd read more about it. So with the Oxford Nuffield coming up, here we are with tons of material . . . Forgive me for using a term that seems derogatory; it isn't meant to be. By 'material' I mean patients' biopsies in my archives that I could take to Oxford and use to further research on the role of human papillomavirus in cervical cancers. I suggested that subject to my professor, James McGee – who had taken the chair at Oxford at the age of 35 – and he accepted the idea. So I embarked on my DPhil and had a wonderful time working in his laboratory.

How exciting was it doing the research?

How shall I summarise it? I wish every South African colleague of mine could have had the opportunity to spend three years in Oxford. This really was science at its best. Here was I, having trained in a developing country where you did hundreds and hundreds of autopsies, and saw thousands and thousands of fascinating tumours and disease processes, now at the cutting edge of science in a First World laboratory. It was just unbelievable.

But there was always this conflict within me, of pulling myself away from the country of my birth and trying to ensure that I got the best out of Oxford. And the cold . . . In those days England was *cold*! And remember, this was coming after a period back home when I was getting used to the idea of having whites as colleagues. In South Africa there was still a sense of animosity between the whites and non-whites, but you'd want to break away from that, because these are now colleagues and you're working in the same environment. You lived in different areas of the suburbs or different doctors' quarters, ate lunches in different places, but you still *worked* together, and there was a level of acceptance and understanding.

And the United Kingdom was now a totally different environment, because they were predominantly Caucasian people. I had to understand: these people are different; these are not the oppressors, the people who voted in the apartheid system. These are now good, loving people and you need to work with them. And, of course, they spoke differently, but the cultural, social, geographical change was an enormously stimulating, exhilarating experience.

You asked me to remind you of February 1990.

Yes . . . Here I was in Oxford, in the bitterness of mid-winter, sitting in a room and watching television. The date was February 11, and here was Nelson Mandela being released . . . In *my* country, and I'm sitting in Oxford with this warm cider, watching this little box. What was I doing in England? My wife and son of three months had not joined me at that time, so I was all alone in England, getting this surreal experience in cervical cancers and human papillomavirus, etc. But the fact was that history was being made back home, and I wasn't there. Can you *believe* the atmosphere that must have prevailed in the country? This was the fruition of my family's contribution to the struggle, and I wasn't part of it.

Did you weep?

Yes, I did.

And did you share it with anybody in Oxford.

No. Who could I have shared it with? I didn't even think about it.

My wife joined me shortly afterwards, and I should tell you, just briefly, this poignant little story, because I hadn't seen my son in over six weeks when they came over. I'd picked them up at the airport with warm jackets, and I placed my son on a counter to put his jumper on. He was asleep. As I was putting it on him he woke up – he was barely five months, and he hadn't seen me for a long time, but he recognised me, reached out and touched my face. My tears just poured down on to his face . . .

So I put his warm clothes on, tears in my eyes, and we bundled ourselves back to Oxford. We lived there for three years, welcomed guests and friends and family, and enjoyed the pubs and the Iffley Lock.

What was going on at home during this time?

We were very conscious of things at home, as conflicts began to emerge between the various factions in the country. The apartheid government, and particularly the police, were still very much in control. The turmoil resulted, in mid-1992, in the Boipatong Massacre, where several people were shot dead. My wife and I thought, 'Oh my God, what's changed? What's going on?' We seriously thought of staying on in the UK at that point. But meanwhile Johannesburg, the University of Witwatersrand, had approached me.

Whilst I was at Natal Medical School, just before coming to Oxford, I was in charge of coordinating the third-year medical curriculum. The professor of microbiology was Jan van den Ende, a dear man. A year after I'd gone to Oxford, he'd taken the job in Johannesburg as the director of the South African Institute for Medical Research [SAIMR], which is a laboratory that was set up historically to service the mining hospitals. His head of department was stepping down, and he called me to interview for the position.

This was the University of the Witwatersrand. Talk about the anxiety of going to Oxford – this was *twice* that anxiety, because this was the doyen of white tertiary education. This was one of *the* two leading institutions in the country – the other was the University of Cape Town – and here I was at 36, a young Indian lad interviewing for the head of department in anatomical pathology at 'Wits'. Several months later they contacted me and said, 'You've got the job.' I was to be one of the first professors of colour in the medical college. All of my consultants in my department were white.

And were you nervous about taking it on?

I was nervous about the idea, not the ability. Not the pathology, not the administration or the teaching – I was comfortable with all of those – but the political component made me anxious. Remember, this was a changing society. You do not just flick the switch and say, 'Okay, yesterday was apartheid, today we have no prejudices.' So I was going to be dealing with people who clearly had prejudices. A simple example: during my tenure at Wits I had people coming to my office, and if my secretary Molly Long had stepped out, they'd say to me, 'Good afternoon, we're looking for Professor Cooper.' Initially that hurt, because here I was, the head of department, and people were asking *me* for Professor Cooper. But eventually I made it into a game, and left it at that. I'd say, 'Oh, he's just stepped out. D'you mind leaving your name and I'll get him to call.'

You were going to tell me more about the SAIMR . . .

The SAIMR is a laboratory system that was created in the early 1900s to provide a pathology service to the mines. Then as hospitals were built in the peripheral areas, the government engaged the SAIMR to take over the pathology of these hospitals too. It grew to be a huge laboratory, doing literally 80% of the pathology of the country. There are various departments within it: microbiology, anatomic pathology, etc. They were academically linked to Wits, and so they functioned as teachers to the university as well as service administrative components to the SAIMR.

So, as professor of anatomic pathology you had two bosses: the dean of the medical school; and the director of the SAIMR, who was your laboratory boss. You answered to the dean about how you were teaching his students and doing research, and for the laboratory boss you took care of service and reporting of patient pathology.

You say you were bedazzled by the facilities at Oxford – how did you take returning to a system that didn't have those kinds of resources?

It was good, and I'll tell you why. In Oxford I was able to meet people that I'd only read about. There was Kevin Gatter and David Mason – these were big names, and I could walk across the corridor and chat with them. You became accustomed to speaking to academics of extremely high calibre: people at the cutting edge of their fields. So I brought all of that experience to Wits and tried to create a similar system. I tried to create a microcosm of it within pathology, which meant allowing consultants and registrars to do research as well as their service work, and to engage in molecular research that I'd learned in Oxford. That worked: we became one of the most successful pathology departments in the country. The way to judge it was at the annual meetings – we made by far the most presentations during the period that I was there, and that brought a lot of joy to me.

So what were you doing? Breaking down academic barriers?

Correct. And creating new things. Remember, this was a changing Wits, and you took advantage of that. Here was this SAIMR director, whom I knew from Durban and who had recruited me. I told him I wanted to create this laboratory and he said, 'Go ahead, Kum. You need money?' He was able to make that happen, and to 'bring in the techs . . .', who were service technicians involved in the daily service loads. He enabled us to bring them into a research laboratory and train them; then he continued to provide a budget so that we could fund research projects that pathologists and trainees could engage in. And that was hugely successful.

We were optimising tests within the research laboratory that we would use for diagnostics for patient care. So it was practical – it enabled us to translate a thought or a question based on observation of some morphological changes into an experiment that you could then perform in a research laboratory. I had consultants and trainees who derived a lot of benefit academically, as well as finding it clinically satisfying.

How much pathology did you find stemming from the mining enterprise?

Oh, a lot. We saw occupational diseases: tuberculosis, fungal infections, diseases that were somehow related to the exposure of patients to the hazardous dusts they breathed. The idea of the link between asbestos exposure and mesotheliomas [tumours in the lining of the chest or abdomen] emerged from the southern African experience, and I had the pleasure of working briefly with one of the original pathologists involved with that.

I just loved the group that I worked with, whom I called 'The Dream Team'. It's a core group of pathologists who still work there, and I still spend time with them.

We've jumped a bit ahead of ourselves. Let's go back to when you first left Oxford . . .

We returned to South Africa late 1992, after three years in Oxford, and I began work at Wits in December of 1992. As a new professor there's a compulsory lecture that you have to deliver to the entire faculty and staff and students. It's the inaugural lecture and it's an open forum. I took the opportunity to pay tribute to my mum for her role in my being there, and that was very poignant for me. I'd achieved, in her eyes, the ultimate in academic pathology in South Africa, as chair of a department, and she was hugely proud. Tears welled in her eyes, and she couldn't stop smiling.

But there's another story I want to tell. You recall my many attempts to get into one of these medical schools, requiring ministerial consent, and getting several letters of refusal? Wits was one of those institutions – and it had turned me down not once, but twice. And I referred to that in the opening deliberations of my inaugural lecture. After paying tribute to my mum, I recalled the difficulties I'd faced in getting to medical school and how I'd applied to Wits and been refused twice. And so I created the motto, 'If you can't get in at the bottom, get in at the top!'

That served two purposes, because, again, I was anxious. One of my senior mentors, on hearing about the job at Wits, said to me, 'Kum, they will chew you up and spit you out.' Those words stuck with me. So here I was, delivering this lecture, thinking: am I going to be chewed up and spat out? And, as I say, the delivery of that little story served two purposes: one, in somehow getting back at the system; but two, it also diffused some of the anxiety, because there was hearty laughter! That allowed me to relax, and I was able to deliver my lecture, which was on a topic close to my heart: the research I'd done in South Africa.

And what about 1994?

April 1994 was, of course, the first all-race election. I'd never voted in my life. My wife and I talked and talked about this, and the day of the actual vote, we impulsively drove all the way from Johannesburg to Durban, to the suburb where we'd built our house before we'd left, and voted there. And it was enormously emotional, hugely emotional – to cast that one ballot! And it just had to be for Nelson Mandela. We came out emotionally relieved, crying. We cast our vote and it changed the way we thought about our country, 'the rainbow nation'.

Then five years later, 1999, South Africa completes the cycle. We were moving in March 1999 to Vermont, and so

I'd sort of missed the voting. The heartache of moving house is difficult enough, but moving country? It's enormously difficult, so voting was not a priority. But then going forward to 2004, we learned that the ANC [African National Congress] government had somehow changed the law so that certain citizens could not vote, and we as South African citizens living in the United States could not.

It caused me so much heartache. Here I was, born in South Africa, and having been through an education system with enormous trauma because of the politicisation of my family. (I wasn't directly involved, so I was worse, I was a wimp: I didn't have the fibre to face up to those security police. But my brothers had done so.) My family were still in the country, but here I was, in 2004, in the United States, and not allowed to vote. I'm telling you that the only time I've voted legitimately in my entire life was one vote in 1994, for Nelson Mandela, and I hold that dear to my heart.

And then his party rejects you.

Correct. And so I'm a citizen of the world, and I refuse to vote again for the rest of my life – for anybody.

So where do you feel you belong?

I don't. I just enjoy my pathology wherever it is.

Two months ago the person who took over from me in Johannesburg, Alan Paterson, was watching television with his family and their house was broken into by six criminals. They were abused. I won't go into detail, but when I think about it, it breaks my heart. Here's a person I had worked with; a family I loved. That's left me drained of feeling a sense of belonging.

My family is my home away from pathology, and pathology is my home away from my family. How else can

I put it? That's it. I try to do my best in whatever I do. So where does one belong? And what does it matter any more?

Okay, the move to Vermont . . .

That begs me to go back a little. In the 1980s as a young consultant in South Africa, I was impressionable, and I saw this name, Christopher Fletcher, appearing on papers with all sorts of dynamic new ideas. As a registrar he was already making his mark in soft tissue pathology.

In 1990 I was in Oxford, and the next Pathological Society meeting was in Belfast. Of course, this was a re-enactment of South Africa, with the guns and the army, the barbed wire and all that. I thought, 'What are you guys worried about? I'm used to it!' Anyway, on the bus to the accommodation, I see this Dr Fletcher sitting at the front. I knew it was him because he'd delivered a paper, and I was impressed by his charisma. But, of course, this is Britain; no one approaches someone on a pedestal! But I went up to him, introduced myself, and we've been friends ever since. We share this passion for pathology, and I love spending time in his department, seeing wonderful cases from all over the world. He is without question the leading soft tissue pathologist presently in the world.

Chris Fletcher obviously knew about what I was doing in South Africa, and when he heard that Vermont was looking for a director he said, 'There's this guy in South Africa you may want to contact.' They called me in April 1998 – I now realise I was looking for a new challenge, that I had done everything I wanted to in Johannesburg – so that's how I came to Vermont. Also I should admit that it was tough living with a family in South Africa: there were social challenges that were emotionally difficult to handle, and crime was a problem. We'd had cars stolen, house broken into. My young son would say to me, 'Are the robbers going to come tonight?' Having your child say this makes you begin to rethink everything: what you've worked for, aspired to . . .

But Vermont was hardly what I thought I was looking for. Vermont's beautiful – you see four seasons, like in England. But it's a country of its own, and we had to adapt to the American way of life. The work environment was second to none, and you were able to function as you did in any department, except that the pathology was different. This was all First World pathology, diseases that pertain to lifestyle – the 'cost of living', as we call it – rather than diseases due to hunger and infection.

So do you find pathology here as challenging as it ever was?

That's a question I've been throwing around in my mind for a while now. I think the answer is that the challenges lie in being in a national position and able to contribute to the way pathology is moving into the new century. For example, I served with the Association of Directors of Anatomic and Surgical Pathology for five or six years, and I've served with the College of American Pathologists ever since I've been in the United States. These national committees help educate pathologists, and help them to define their role in their own laboratories. I think that's the role I see myself serving more and more – that's where the challenges lie.

You say, 'moving into the new century' – do you think pathology is changing?

Yes, it is. I don't know what it's changing into, but what we are leaving behind is classical pathology – and in my own discipline, that's the classical surgical pathology that I learnt from the English teachers who influenced me in Natal

Medical School. Whether it's for the good, or not-so-good, time will tell.

You want to leave the pathology world a better place, and I think that's where some of the gloom sets in. It's difficult to predict where we're going, especially bearing in mind the wide disparities between countries. In Africa, for instance, they're having difficulty getting basic stains; in the United States it's whether to use the super-stain or not. You cannot say these disparities don't affect us all – this is a global village. So where do we go from here? I really don't know. My immediate focus is to take it a day at a time, and contribute whatever I can.

SHAKEN BABIES AND UNSHAKABLE MINDS

Waney Squier
*Consultant Paediatric Neuropathologist, John Radcliffe
Hospital, Oxford*

Waney Squier went into pathology after finding it hard, as a young paediatrician working on the wards, to distance herself sufficiently from the suffering of her young patients and their families. Pathology, she thought, would give her the chance to indulge her passion for medicine – and particularly for teasing out how disease happens – without having to confront personal tragedy head on. Yet it hasn't always worked out that way: in the aftermath of Alder Hey, she found herself having to justify the practices of her department to individual, extremely angry and upset parents.

As a paediatric neuropathologist, Squier is particularly interested in brain development, and with what can go wrong with the process and why. Recently she has become involved in the controversy surrounding shaken baby syndrome, questioning the scientific evidence for the original hypothesis. Her challenge to received wisdom has put her at odds with some of her clinical colleagues, and involved frequent appearances in court as an expert witness – an often extremely adversarial experience, given the very different mindset of lawyers and scientists. 'That's the problem with court: scientific fact is lost,' she says.

'The lawyers are there to win the case, and that's all they want to do.'

———•———

I got involved in the shaken baby debate after reading the work that Jennian Geddes had done in 2000 when she raised the issue. She did a *fantastic* neuropathological study. She looked at 50 cases of children who were thought to be non-accidentally injured. It was the biggest series that had ever been published, and it was a very good descriptive study showing that most of these babies don't have the markers in the brain that we would use to diagnose trauma in adult brains. All they have is swelling and lack of blood supply to the brain. And she started asking questions: 'How many of them are trauma? Is it the trauma that's causing damage to the brain stem, and that then causes the child to stop breathing, and the brain swells as a downstream effect?' Then she suggested that maybe the blood is coming from the dura itself [the outermost and toughest of the three membranes covering the brain and spinal cord], rather than from the torn blood vessels. She was the first person to suggest that, and she started saying, 'Maybe we've been getting it wrong.' And I sat up and thought, 'Well maybe I have, too,' because I had simply believed that finding subdural haemorrhage, retinal haemorrhage and a swollen brain meant shaken baby syndrome.

Had you seen quite a number of such cases yourself?

I had seen some, and I'd actually diagnosed them as shaken baby syndrome. It rather frightened me that I'd just accepted what I was told, so I started looking at the literature. When I started reading what was actually written about this syndrome, I realised that the whole basis for it was incredibly insecure. Then I got into discussions with Jennian and various other people who were questioning this, including a lively and informative group of forensic pathologists, biomechanical engineers, surgeons and radiologists, predominantly in the United States. We exchange letters and comments about cases, and discuss and question everything that's been written.

It's actually rather sad to admit that, at my advanced years, this is the first time I've become absolutely stringent and rigorous in my examination of any scientific paper, to see that the methodology is good enough to accept the findings. I was, of course, trained to be careful, but now I'm so much more vigilant, and particularly in this area. So much of what we're told we accept, and we don't ask enough questions.

Somebody recently quoted one of his lecturers at medical school, who welcomed them by saying, 'You'll be given a lot of information here, and probably 50% of it will be proved, subsequently, to be wrong. The only problem is we don't know which half will be wrong and which right!' [*laughs*] So we all have to keep asking questions. I certainly became aware, when I started asking questions about this, that in the area of shaken baby syndrome the whole diagnosis depends on, 'We believe it; we've seen it before; we know,' but not, 'What are the facts to support our belief?'

So if shaken baby syndrome has been misunderstood and perhaps misdiagnosed for such a long time, surely there must be quite a controversy?

There's a *huge* controversy. It frightens me because it's not just making the wrong diagnosis; it's actually wrecking people's lives. I think my worst moment in this whole saga came when John Sweeney interviewed me for a BBC television programme. I was warned about five minutes before that he was going to ask about a case I'd diagnosed a couple of years earlier, and I looked at my report and it said: shaken baby syndrome. He then set up the interview, sat me down, handed me this report and said, 'Two or three years ago . . . you diagnosed shaken baby syndrome. Lorraine Harris has been in prison for three years. Do you still believe in shaken baby syndrome?' I said, 'No, I don't.' And he said, 'Well, what do you think of your part in her conviction?' I was on camera; I was absolutely devastated. It was a good piece of television. I said, 'I'm just horrified. Because now I've had a chance to rethink and I don't believe I'd make that diagnosis again.'

In fact, as a result of that interview, I was asked to appear in the Court of Appeal in 2005 when her case was heard. Her conviction was quashed, and the situation really went home to me when her solicitor, a delightful man called Campbell Malone, had to ring her and say, 'Lorraine, your conviction was wrong.' Firstly, when the baby died she was taken to prison, and she wasn't allowed to go to his funeral. Her five-year-old child was taken away and adopted while she was in prison, and when she came out she was only allowed access twice a year. Her husband had left her and her parents had died . . . Her baby died of a natural disease, and yet the system has wrecked her life about as comprehensively as it's possible to do.

So how did you come to terms with that?

It was devastating. I've tried to come to terms with it by saying that I'm going to look into every detail and put everything I can into each of these cases.

Has it frightened you about making diagnoses, or have you recognised that medicine moves on and it was an honest mistake at the time?

Yes, it was an honest mistake at the time. Except that, had I had more rigorous criteria for believing what I'm told, maybe I wouldn't have made that mistake in the first place. But, you know, we've all made mistakes, and we've all moved on. But it still worries me, because the amount of information we have is often insufficient to be absolutely certain of a diagnosis of non-accidental injury. Often the best you can say is, 'I really don't know.' And that isn't helpful to lawyers. The police hate it. But if it's honest, then it's what we have to say.

What's made me really passionate about this is what's at stake for families, as the Lorraine Harris case demonstrated. I've since had a number of cases where families have been found 'not guilty' eventually, but the suffering they go through in the process is just awful.

There was another case that really upset me. This baby had been in hospital for only 44 hours before he died. That is not a long time to establish guilt. In the notes the nurse had written, 'Baby E is on a ventilator. Parents are being interviewed by the police at present. The police have said that if his condition deteriorates they will bring them back to the hospital.' So the parents were deprived of the last 44 hours of their baby's life – they were in the police station being accused of injuring him. Just imagine, as a mother, how that must feel. And they were found innocent. Their case was thrown out of the family court. Those parents were deprived of the last 44 hours with their dying baby; you can't ever give that back to them.

Then there is the polarisation of medical experts. There are the people who believe that babies are injured by families . . . Well, we all believe that babies are injured by parents and carers, but some people are so passionate in their belief in child abuse that they won't brook any questioning. I'm regarded as off the wall, a bit of a maverick, because I keep saying, 'I need to see some evidence before I accept it,' and so we get into a very difficult, adversarial situation, with the hawks and the doves.

And how do you manage it? Is it in your nature to fight for causes?

I'm absolutely not confrontational! This is part of the reason I'm a pathologist – because I'm not good at dealing with dying people, or clinical situations where you have to make decisions immediately: do you give this drug or that drug? No, I'd rather go away and think about things quietly. But this issue has actually fired me up. A few years ago if you'd said, 'Go and be a witness in court,' I'd have said, 'No, no. A clever lawyer will talk me into saying black is white and I couldn't possibly stand up to that.' But I've become so involved in this whole area, and I'm so up to speed on the evidence we have, that I'm willing to go into the witness box. Because I know that this evidence has to be put before the court, otherwise there are going to be dreadful miscarriages of justice.

Are you creating a stronger evidence base for alternative mechanisms for this condition now?

Oh, this is a terrible problem – because, yes, we've got quite a lot of cases now that we should be writing up. But every time I'm just about to write up a paper another case comes . . . 'In court next week; we need your opinion.' It's really hard to get time to do it all.

Interestingly, perhaps a sign of the way this is going is that I was invited to write a review on the topic of subdural haemorrhages for a paediatric journal. I wrote my paper. It was reviewed by two referees for the journal, who rejected it saying, 'It depends on old literature. It's not the current belief. This is a dangerous piece of writing.' It was, as far as I was concerned, just an analysis of what we know about subdural haemorrhage using a lot of data I got from the old literature, and *questioning* the standard belief. It's not as if it's old technology that we know was faulty or has been superseded. This is basic, old, observational pathology: just simply people describing large numbers of cases – more than we can study today. That doesn't change.

So how common is shaken baby syndrome?

Oh, I probably see a new case every week. In fact, just this morning two policemen brought me a case and I said to them, 'I don't think it's shaking; there's no evidence. The baby doesn't have a broken neck; there are no grip marks. This baby may well have been impacted – either dropped, or smacked across the head or thrown against something.' But they kept saying, 'Yes, well if the father hadn't done it . . .' And they were making this shaking gesture; they just can't get away from the idea that 'this is what families do to babies'.

So we've got a long way to go to get that out of common parlance, because once you've got shaking, you've got 'intentional injury', whereas if you've got impact, it can be just as intentional, but it can also be accidental. So you are at least putting things on a level playing field. Then you can put your facts before the court and the jury can decide on the evidence. But once you've got shaking, it's intentional and it's dead easy; it's a downhill run to prison from there.

As the pathologist, aren't you in a powerful position to say, 'This is not shaking'?

I do say that. So we have to look further, to demonstrate impact or prove that there was some other sort of neglect or abuse. The detail is so important – in some of these cases we find a huge subdural haemorrhage during post-mortem. Well, that goes with shaken baby syndrome. But then you go back to the scan that was taken the day the baby was admitted to hospital, soon after the collapse, and there's no subdural. A scan two days later and there's a bit of bleeding; then you come to post-mortem and there's a lot of bleeding. The hawks will grab this and say, 'Well, subdural haemorrhage – it must have been shaken.' But then I ask, 'What about the scans?' What's happened, I think, is that the blood is coming from somewhere else and it's oozing; the baby's sick and it's on a ventilator; the liver's not working; blood clotting is not working; and this blood is collecting over a period of days.

In fact, I've got to go to court tomorrow on a case where exactly this has happened. The prosecution is running a diagnosis of shaken baby syndrome – and I'm on the prosecution side! I'm a real thorn in their flesh, because I'm saying, 'You can't say it's shaken baby syndrome. There's no broken neck; there are no bruises where the baby's been gripped, and this is a huge baby.' And the blood would have been there on the first scan, but it wasn't. The radiologists on both sides who've looked at this case have said, 'There's no blood on the first scan.'

Do they agree with you that there are bigger questions?

I don't know. I'm saying, 'I think this baby's probably had impact injury on separate occasions.' But that's not very good for the police; they want it to be shaken, and they want the injury to have happened 'at 12 o'clock in the morning, 3 November 2003', or whatever, because there was only one person with the baby then, and it makes it really easy for them to identify a perpetrator.

Because 'impact' is a much more complicated story? It could have fallen off a high chair or something?

Yes. Particularly if you say, 'Well, I think it's been impacted on a number of occasions, because there's old bleeding as well.' Then they say, 'That's really difficult because the parents have been looking after this baby, and there's a baby minder, and aunts and uncles . . . We need a nice tight story that fits with the baby collapsing in the presence of only one person.' And that's where it becomes very difficult, because my colleagues run with the story of shaken baby syndrome, and I think we've got to look at other things.

I go to meetings here at the hospital very regularly where I will be greeted with, 'Is it another "killer milk syndrome", Waney?' – because I believe that some little babies choke on their feeds and die, especially if they've had a difficult delivery and they've had bleeding at birth. That bleeding may not have completely healed and it may re-bleed. And if they choke they're going to raise the pressure in their heads, and they'll get another subdural bleed.

Is this the first time you've found yourself challenging the orthodoxy, or does being a pathologist mean you're always in a slightly challenging role?

No, not at all. We're quiet people who go away, work in the back room and send out the diagnoses. No, I'm absolutely not a confrontational person. But I've got to the stage now where I feel so strongly about this that I think, 'I've only got a few more years to go. I can be difficult; I've nothing to lose. But there is a lot to be gained if we can prevent miscarriages of justice.' So I'm willing to be as controversial as it takes – *if* I've got the evidence. I work on the facts in front of me.

Building up the evidence, of course, means retaining tissues from autopsy, to study and to archive. How good are you at broaching this issue with families?

Well, I'm actually quite impressed with myself! I thought it would be terrifying. But going through the whole Alder Hey thing – having very angry, grieving parents storming up to the hospital saying, 'You stole my baby's brain; how could you do it, you people?' Then having had a few that I've sat down with who've gone away saying, 'Oh, now we understand. You learnt something; his little life wasn't in vain' – that really gave me confidence to believe that sometimes I was getting it right with families. So now I'm quite willing to do it, and I often tell the paediatricians or the coroner's officers they don't need to worry about talking to families about autopsies and brain examination – I'll speak to them.

Tell me a bit about your own personal path into pathology. What kind of family did you grow up in?

My family was completely non-medical. At school I really enjoyed biology, and learning how tissues work, so I wanted to do zoology and do research. My brother, who is seven years older than I am, perhaps helped me. When he was doing his zoology A level, he used to make me revise with him, ask him questions and test his knowledge. He did natural sciences at Cambridge and then went into research in a dental school in London. And he soon realised that there's a very real disadvantage to being a pure scientist as opposed to a dentist. If you're a dentist you can go out on a Saturday morning, do a few fillings and buy yourself a nice smart car. But if you're a scientist in this country you're on a very different salary, you're on this constant search for funding and you don't have tenure. So he ended up going to the States, where it's much easier to work as an academic scientist in a clinical environment.

Anyway, the upshot was, when I was applying for university, he said, 'You're going to do medicine.' And I said,

'No, I'm absolutely not. I couldn't possibly scrape people up off the M40 and put them back together again.' But he said, 'Do it, because you'll find that when you go through medical school you'll have opportunities to do a huge range of things. You can go and do research, but just get yourself qualified in medicine.' And so I did.

And I loved it. Leeds was a great place, because we had a huge number of patients per student. Being relatively few medical students in Yorkshire we had lots of opportunities to go to small peripheral hospitals – to do paediatrics in Bradford, psychiatry out in Killingbeck, or wherever. Compared with my friends in London I think we were very lucky because we got tremendous clinical experience.

At the end I decided I was going to do either paediatrics or pathology. I was still fascinated by what goes on at the tissue level, but I wanted to try paediatrics because I enjoyed it. I thought kids were tremendous fun, and you couldn't have a stuffy old professor shambling around in paediatric wards because the kids wouldn't allow it. So I did paediatrics for a few years. But then I found that I'm better off not making rapid decisions and not having to deal with the sort of emotional environment of sick and dying children.

I always had to take a deep breath before I walked into casualty, not knowing what I was going to see and with nurses saying, 'Right, Doctor, here's a child having seizures; do something.' I'm not that sort of person. Even now with my diagnoses I never make a decision at once. So although I really enjoyed paediatrics, I need time to think – and I get a bit crabby when I've been up all night! So I ended up going into pathology.

And how did you decide on neuropathology?

I was fascinated by the brain and brain development when I was a student, and found neurology quite interesting. But you never get any answers in neurology: you do a few tests and then you sort of guess; whereas in neuropathology you've got a really complex organ and a lot of very difficult questions to answer. I think it was that that interested me. And having been interested in doing paediatrics, I thought that looking at the developing brain would be interesting.

And, of course, it's completely fascinating. Much of adult neuropathology has to do with dementia, Alzheimer's disease, strokes, and things that really are on the way downhill, whereas in development every *week* the brain is different. It's like putting a puzzle together. You look at a brain scan of a six-year-old child who's got cerebral palsy, for example, and you say, 'How can we tell what happened?' Well, we know that this structure has already formed, and we know it forms between 11 and 17 weeks of gestation, so the event must have happened after 17 weeks. Okay, we know that neuronal migration has been interfered with as well, because there's malformation there in the cortex, and that happens before 23 weeks, therefore this must have been between 17 and 23 weeks' gestation. So you start asking questions: what happened at that point in time to cause this pathology? Nobody knows. So you go back into the notes and you find, hey, this mother was admitted to hospital after a road traffic accident; she was kept in a couple of nights, and there was abdominal bruising. Wow! That might be your answer. You're doing this detective work that can really only be done on the developing brain, so it's absolutely fascinating to me.

What about home life – do you take your work home?

Oh, that's got worse and worse! My two daughters have now left home, so I've got lots of time at home to work if I need to. And the medico-legal cases, of course, are done outside my NHS hours; there's an awful lot of paperwork, so the dining room is stacked up with papers.

How did you manage motherhood and your job?

[*Laughs*] That's quite interesting, because when my daughters were little I was just starting as a consultant in neuropathology. They were at the 18 months and just-born stage, and I was driving into Oxford each day, dropping them off at nursery and driving back home. I was working part time then, and I would never advise any woman to go part time, because when you're not there the work just gets plonked on your desk. You get to do the same amount of work as everybody else but in your reduced hours – and with reduced pay. So at that stage I was just fraught most of the time. The girls have said to me, 'Those years were absolute hell; you were so *busy* all the time. And you were always bad tempered!'

Did you feel guilty at the time?

Oh yes! Constantly guilty. Even though one's paid half time, I was aware that I would be leaving at 3.30 p.m. when the nursery closed, and my colleagues would be looking at the clock as I left. They probably just thought: oh, it must be tea time; she's off. But I was thinking: they're checking that I'm going. You compensate for that by working harder than anybody else. And you're guilty because you're not always there for your children. And at weekends, you know you should be doing things with them but you just actually have to prepare for a lecture . . .

But on balance, looking back . . . I find the work so fulfilling now, and it is lovely occasionally when you look at something and think: I know what it is because I've seen it before. Also, now I find I'm confident enough to take on the battles of shaken baby syndrome, and I don't think I could have done that if I'd given up my career and then come back. So it has been immensely rewarding. And I do think that for all the girls' feeling they had a rough few years, they are actually now incredibly proud, and pleased to know that they've left me with a fulfilling life, rather than a huge empty nest and a less-than-satisfying career.

Let's go back to your own childhood . . .

My father went to an excellent school and wanted to be a farmer. But his father was killed in the First World War, he had two brothers and his mother said, 'You have to go and work. Your uncle can get you a job in the bank.' So he became a bank manager and really didn't enjoy it. But he retired when he was about 58 and bought a smallholding. He had a house in the Cotswolds, a cow, a calf, a dog and a stray cat, some ducks and chickens, and he was so happy for 20 years, it was absolutely lovely.

My mother had been a secretary, and she was full time at home and was terribly ambitious on my behalf. She didn't want me to be stuck at home not having a career. She was very keen for me to go to university and get a decent job. There were times when I thought her approach wasn't quite balanced. When I was expecting my first baby it was, 'Isn't this going to interfere with your career?' And when I told her I was pregnant the second time, 'Oh, not another baby!' I desperately wanted children and I thought: I'm going to do both; and fortunately I did. I think my mother was influential in making me want to do well; my father was much more laid back.

So who have been the people you've turned to for inspiration and support in your career, especially with your challenge to the shaken baby orthodoxy?

That's an interesting question. My daughters have always been very supportive, but I'm terribly careful about not making them feel I'm depending on them. So colleagues, other people making these diagnoses. Some of the lawyers I've worked with will be really supportive in saying, 'Yes, you've got to say what you think.' So, too, are other professionals who think as I do, particularly in the United States. They know what it's like to be really roughed up in court, as it were, and if you've had a bad day you email them and they'll say, 'Yeah, tough . . .'

In fact, I don't think we get nearly the maltreatment they get in the States – personal character assassination is very much on the cards there. It's not as bad yet over here, but I was in a case a few months ago, and only recently I read the transcript of one of my colleagues being cross-examined in the witness box and being asked, 'D'you agree with Dr Squier or not?' He said, 'Yes I do.' They said, 'But she's a lone voice; she's going against the orthodox opinion, isn't she?' *I* wasn't asked those questions, but my colleagues were. This was clearly the prosecution building up the case to set me aside as completely off the wall. I mean there are lots of people who believe what I believe, but in this case that was his tactic – to get the jury to believe that whatever I said was not to be taken seriously.

One of the things that's so important in court is the way you come across to a jury – which is bad news really, because a jury will see you as an actor and will like the way you are or not, and that will probably be more important than what you're saying. One has to realise that this isn't pure science; this is theatre as well, so it's quite tough.

The other ridiculous thing is that what I take to court is what I see down the microscope – my findings in a particular case, which is intensely complicated, detailed pathology – but that has to be translated into 'Did this person do it or not?' You're allowed to answer yes or no. You never get the opportunity in court to say, 'Can I just give you a 20-minute rundown on what subdural haemorrhage is? Can I explain what we think about this?' The judges won't want to know, the lawyers won't understand it, and they haven't got time.

In 2005 the Court of Appeal looked at four cases of shaken baby syndrome, and they judged the new hypothesis proposed by Jennian Geddes on the alternative source of subdural bleeding – from the dura itself rather than from the torn blood vessels. It was a complete farce to see lawyers struggling to understand *incredibly* complex physiological explanations for what's happening in baby brains; it's hard enough for us to understand, and we've been doing it for years. But for lawyers to try and understand it, and then to say, 'Well, the Court of Appeal found that the hypothesis didn't stand up to scrutiny,' is a nonsense. And that's the problem with court: scientific fact is lost. The lawyers are there to win the case, and that's all they want to do. So it is very frustrating.

What would you say, looking back over your career, have been your main motivations? Is it the puzzles of the science, or working for the families, or what?

Working for the families. If you do something that actually makes a family feel better when a child has died, and a family writes and says, 'Thank you. You gave us a lot of help,' that is tremendous. Because, as pathologists, we don't meet families. We don't get bottles of sherry at Christmas! So the odd letter from a family is hugely gratifying.

But what really, I think, motivates me is having all of this material to look at. It's such a *privilege* to have all of this pathology that we don't understand, so many questions to answer. We don't need to do nasty things to rats and to monkeys in cages, because we've got so much material that comes to us as part of our diagnostic work. And we can use that to learn. Every case that somebody sends me, I know that there's going to be something that I'll learn from it. So that's what keeps me stimulated and keeps me going.

Okay – one final question: how do you relax?
I listen to music, go to lots of concerts. And I drink! [*laughs*] I'm very interested in wine and do a lot of wine tasting. I have a nice circle of friends in Oxford and we're all passionate about wine, so it's not just a question of blotting everything out, it's a reasonably academic pursuit of wine. And that, of course, is something that really takes you away from your day's work, sinking a few bottles of wine with good friends.

DIAGNOSIS: A MARRIAGE OF ART AND SCIENCE

David Levison
Professor of Pathology, Dundee University

Many of the turnings David Levison has taken in his professional life – from opting for medicine in the first place, to the areas of expertise he has developed as a diagnostician, and the research projects he has taken on – have been serendipitous. 'But I think that's the way you enjoy life – you have to make use of the opportunities as they come along,' he says. Having got 4% in his first pathology exam, he never dreamed this would eventually become his great interest. Levison was thinking of becoming a GP, but when he started as a junior doctor on the wards he felt wanting as a diagnostician, and believed he would do better if he had more understanding of pathological processes. He took what he thought would be a short-term position in pathology and has never looked back. Key lessons he's learned, he says, are that 'doing things properly and having an open mind are tremendously important'.

Disturbed by the attack on pathology following the Alder Hey controversy, Levison has also become involved in the politics of medical research, as part of a group that advises the Scottish Parliament on relevant legislation. He believes there should be 'presumed consent' that any tissue taken for diagnosis can be used also for research.

Over his career, Levison has seen dramatic changes in pathology practice and has embraced with enthusiasm all kinds of new tools as they became available. But when his young wife died suddenly, it was the oldest technology in the book – the autopsy – that was most useful in establishing what had happened and bringing closure for her family.

———•———

Was there a moment when you suddenly realised: 'Pathology is what I want to do'?

Yes! It was in my second year in pathology and I remember it precisely, because the GP's practice in Carradale on the Mull of Kintyre came up. That's a place I like, on the west coast of Scotland. I'd been there and thought, 'Now if that GP practice ever comes up, I'd quite like that.' And when it came up I thought, 'Am I going to apply for it? No, I'm actually really enjoying what I'm doing now.'

I'd got interested in a post-mortem case, actually. We'd seen this lady who'd died of endocarditis – infection of the heart valves. It looked a bit unusual, with big vegetations on the valves, so we'd sent stuff off for bacteriology and done detailed microscopy looking for organisms. It turned out to be due to psittacosis, which is a disease you catch from birds. We discovered that this woman's husband kept exotic tropical birds, and she used to go in and sweep out the cages.

But this case also brought to mind another case from three years earlier. The senior pathologist at the time, Bill Guthrie, had done a post-mortem, and I'd remembered it. I was a medical student, and we'd seen this chap on the ward. The doctors thought he had psittacosis, and they thought there was something wrong with his heart. Then he died.

A couple of us had been interested in the case, so we went along to watch the post-mortem, but we were shooed away: 'No, you can't come in – danger of infection.' I remembered that case well, and the consultant supervising me on this second psittacosis case was again Bill Guthrie. We put our two cases together and wrote a paper for *The Lancet*. That was my first introduction to research and writing a paper.

I was involved in that when the GP practice came up in Carradale, and I thought, 'Well, there's a big difference between being a GP and being an academic pathologist, but gosh this is interesting. Yes, this is what I want to do.'

Were you then able to go back through historical samples and find out whether psittacosis was more common than doctors understood at that stage?

Well, that did sort of open up the area. But at that time we didn't have molecular techniques. All you could do was, with the fresh tissue, try to grow the organism. So I didn't have the facilities or the tools to take it forward. Nor the opportunity – it's a pretty rare disease. No, I got on to other things after that. But it taught me a lot about how exciting and interesting things could be. And also how doing your job properly – just being careful, and thinking about it – you get a lot more out of it than you expect to.

Is keeping an open mind important, because I should imagine it might be easy to miss things if you have preconceptions?

Yes. I think doing things properly and having an open mind are tremendously important. That again . . . Can I brag about another diagnostic coup?

Do, please!

It was my first medical house job and I was in Leeds General Infirmary. A lady came in for coronary artery angiography. She was in her mid-thirties, I think, and she'd had a number of attacks of heart failure, so she'd come in to have angiography to see what was wrong with her coronary arteries. I talked to her, clerked her in and examined her. Everything was normal: blood pressure and everything else. And I just took the time afterwards – I probably hadn't anything better to do! – I sat down and read through her notes. I noticed that each time she'd been into hospital ill, when they'd taken her blood pressure it had been high, but when she was seen as an out-patient, and when I'd clerked her in this time, her blood pressure was quite normal. Having been reasonably well taught in pathology, I thought, 'I wonder if she's got a phaeochromocytoma,' which is a tumour of the adrenal medulla that produces adrenaline and noradrenaline. So I thought, 'I'll just send off urine to see if the relevant hormone level is raised.' It came back in the clouds!

I was terribly popular with my consultant there, the cardiologist, because this lady had been under a cardiological team in another hospital that he wasn't involved in and they'd not picked it up; then she'd come to him and his houseman had picked it up! And it was very important . . . She had radiology of the abdomen, and there was the tumour. It's usually a pretty benign tumour. She had it removed and was *totally* cured – and didn't have to have the risky angiography.

I tell that story to the students just as they're graduating. I tell them, 'It's not being intellectually brilliant that will make you effective all of the time, but just doing the job properly, and taking time to look at the background and so on. You can sometimes get amazing things out of it.'

Let's go back to your roots – what sort of family did you grow up in?

I was born in Perth, Scotland. My dad was a parish minister; he's still alive, and he's one of my heroes. I remember my childhood being very happy. The other thing I remember is my mother's asthma – summer days with her in bed and the windows all open and her gasping for breath, and adrenaline keeping her alive in several acute attacks. She was a chronic invalid until I was in my early twenties, and then Becotide inhalers came along and . . . [*snaps his fingers*]

She was better?

Fantastic! I mean, asthma comes and goes, and she wasn't in bed all the time, though my childhood memories have her there a lot. But when I think about going on holidays up to the north-west, she was always there and able to come on walks with us and so on. I remember my childhood with fondness.

Why is your dad one of your heroes?

He was obviously well respected in his job, respected by his parishioners, and that was apparent just in going around the town. And he was easy to talk to.

He always used to take four weeks' holiday in the summer, and that was the highlight of my year as a child. We would go up to the north-west of Scotland, take a house in places like Arisaig or Mull. Then he got into caravanning in 1951, so it was caravanning from then on. He used to pull that caravan with a 1936 Austin Ten, which wasn't nearly strong enough, and I used to sit in the back of the car holding a brick. We would get stuck on hills regularly, and my job was to jump out of the car, run round and stick the brick under the wheel so we didn't run back downhill. [*laughs*]

My dad had an old canoe from his university days that he used to bring on holiday, and he also introduced me to sailing . . . Some of the not-very-safe things we did in boats!

So how did your career in pathology and your research interests evolve?

A significant thing happened during my training in Dundee, I remember. This was also on the diagnostic side. We got a professor of medicine who started doing renal [kidney] biopsies, which we'd never had before in the pathology department. One of the consultants was dealing with them, and he was using the traditional stains that we used in the department, and showing the biopsies to me as he went along. I read up a bit about how you did renal biopsies, and I found that everyone else was using a different silver stain from the one we were using. So I said, 'Look, we should be doing this other stain because you can see these little spikes that help you to diagnose particular types of renal disease.' He said, 'No, I'm used to this stain,' and he persisted with the silver stain we were used to in the lab. I must have been a really irritating person to have around, because I said, 'Look, I don't think we're doing a very good job here.' He just *glowered* at me, picked up all the slides, dumped them by my microscope and said, '*You* bloody well report it, then.' So I did. [*we laugh*]

So after that, even though I was a junior person with no qualifications, I became the 'expert' renal pathologist in the department! But that was good, because it meant I had a niche. And that was very useful when I moved to Barts to my first senior lecturer job, because I knew I could do kidneys, and they didn't have anyone there at the time. The chap who'd left had done all the kidneys, so I had a really good niche and was able to build from that.

Another great opportunity I got when training in Dundee was to learn gastrointestinal [GI] pathology. We had a new professor of medicine who was a gastroenterologist, so we started getting endoscopic biopsies. This was just the time, the late 1970s, when GI endoscopic biopsies were coming in, and again that was very useful when I went to Barts, because there was nobody picking up the GI stuff. So even though I'd been number two in Dundee, I was number one in Barts. And I learnt it very quickly, because I had to.

But weren't gastrointestinal problems a major area for diagnosis? What were they doing at Barts before you arrived?

Before endoscopy was developed such diseases could not be biopsied. Our understanding of gastrointestinal pathology was based on surgery and what came out of that, because we'd really only ever seen stomachs that had been removed because they had cancer or because they had an established ulcer. We never got to see any of the lead-up pathology, or the pathology from people who had lesser symptoms, or pre-cancerous or pre-ulcerous disease. So really, when gastrointestinal biopsies became possible one had to learn it from scratch. I mean, one felt a bit at sea, but we were all learning together.

I enjoyed my time in London very much indeed, professionally and socially. I was working with fantastic people. I got to know Basil Morson, who worked in St Mark's Hospital just up the road from Barts. He was on his own, and we became very friendly; when he went on holiday I would do his work for him. He wrote *Morson and Dawson's Gastrointestinal Pathology*, which was really the first decent textbook of GI pathology. It was based on very careful studies of surgically removed specimens, and then on the biopsy specimens also. It was fantastic to have the opportunity to work with somebody like him, and learn from him.

This was obviously virgin territory – when you first saw these biopsies, how easy were they to interpret?

At first it was very difficult to work out what might be the significance, if any, of what one was looking at. And I know that I – along with virtually every other pathologist – dismissed the little things that turned out to be *Helicobacter pylori*, the main cause of ulcers and various other stomach diseases. Robin Warren, who won the Nobel Prize for Medicine, was the pathologist who wouldn't let go – he just kept on at his physicians about what he was seeing, saying, 'Look, I'm sure these are bacteria.' And everyone said, '*Pah*, you don't get bacteria in the stomach . . . *Acid!* How could you possibly have bacteria in the stomach?' And he said, 'No, I'm sure these must be bacteria . . .' Then, you know the story, Barry Marshall [joint winner of the Nobel], with nothing better to do, went and talked to him and got persuaded. Then they worked together, and the rest is history.

And once Warren and Marshall started publishing, did you go back and have another look . . . ?

[*Laughing*] Oh yes, the bacteria are easy to see.

Why had you missed them?

Because I believed the dogma of the time – that there *weren't* bacteria in the stomach. The alternative explanation was that they were just bits of sloughed cells. I mean, Warren and Marshall were pilloried! But they hung in there, and it was the established view that was shown to be stupid and wrong in the end.

You've been talking about the fascination of diagnosis and microscopy – how conscious are you of the patient behind the slides? Are you more of a scientist than you are a physician?

[*Pause*] I'm both. If I'm diagnosing, then I'm always very conscious there's a patient at the end of it. I don't like writing a report where I have to say, 'I can't make up my mind,' though I have to fairly often. And I don't like writing a report that says, 'To make this decision you're going to have to go back and get more material,' and I have to do that fairly often too. But I hardly ever see patients. And I don't miss that at all. I have to deal with my colleagues, who are just as difficult as patients. [*laughs*]

Yes, I've been interested to hear that there's a stigma attached to pathology, even sometimes amongst others in the medical profession. What do you feel the status of pathology is?

I think it's a lot better than it was, with the emphasis now being on multidisciplinary teams trying to collate all the evidence and with the pathologist always being part of such teams. And you can't get away from it that it's often the information that the pathologist provides that is the crucial thing. Not always. Okay, with imaging techniques getting better and better, they can often see the extent of a disease process even better than we can. But they're still only looking at grey and black and so on . . .

And don't you still need an explanation of what you're seeing with imaging techniques?

That's right. I think it's what excited me about pathology – the fact that you can actually start to understand the *mechanisms* of things. And I think microscopes are terrific things for that.

One of the reasons I like pathology is that it provides more answers than many other branches of medicine. But it is by no means a definite science. I mean, it's *all* judgement. 'Why do you say that's cancer when you look at that slide?' 'Well, because my experience of looking at this sort of thing before is that a patient who has this will be dead within a few weeks if you don't do anything.' It's based on experience rather than anything more solid than that.

In your time, you've seen the advent of some major advances in pathology, like immunohistochemistry in the 1980s – how big has that been?

It's been *huge*. This technique allows you to visualise where a particular protein is in a cell or in a section. You use antibodies that are 'labelled' with something you can see down a microscope as a stain, so, if you have an antibody to protein A, you know that wherever you see that stain, protein A is there. It was the advent of monoclonal antibodies that are absolutely specific to particular proteins – i.e. just one antibody to one protein – that has been very helpful.

Before we had immunohistochemistry, if we saw a mass of malignant cells we could tell it was a tumour, but not definitely if it was a lymphoma, which is curable; or if it was an undifferentiated carcinoma, which virtually nothing will touch; or a sarcoma, for which there might be some specific treatment. Now, with immunohistochemistry – and the molecular techniques as well – we can tell in almost every case, 'Yes, that's a B-cell lymphoma, usually curable. Ninety per cent of people with this sort of tumour will still be alive in five years, *if* they are treated with this particular regime.'

*How can you tell? Because it shows you which proteins are
there and where they will have come from in the body?*

Yes. If the malignant cells are expressing lymphoid antigens,
then you know it's a lymphoma. If you can tell they're *not*
producing lymphoid antigens but are producing proteins
you usually find in epithelial cells, you can say, 'That is a
carcinoma.' Before this technology became available, I could
only make an educated guess.

Tell me about the study group you have here.

I'm chairman of the Tissue Bank Committee here, and the
principal grant holder for the Tissue Bank. We're funded by
Cancer Research UK, and we collect tissue that's surplus to
diagnostic requirements – we have to have signed written
consent from the patients to do that – and I chair the
committee that deals with requests to use that tissue. So I'm
getting quite involved with the political side of things, and
am on a group that is advising the Scottish government on
exactly what the legislation should be for research tissue in
Scotland.

I'm going to play very hard in these negotiations for the
'presumed consent for the use of surplus tissue for research'
position. If I get batted back from that, then I'll go for the
position that if somebody's having an operation under the
NHS they should *always* be asked, at the same time as they're
being asked for consent to the operation, if the surplus tissue
that the pathologist doesn't require for diagnosis can be used
for research. If we can achieve that, it will be a significant
boost to tissue research.

*How inhibiting are the rules at the moment? Is there a brake
on medical progress?*

I think so. I know of studies that have not been done because it's just not worth the effort of going through the ethical hoops to get permission. It's hard enough to get money to do research, but when you've got to do this in addition . . . I'm sure I wouldn't have done a lot of my research projects. I might not have gone into academic pathology at all had you to go through the rigmarole then that you do now. It really does slow things up.

Lesley Christie, who's a clinical research fellow here, has got funding for herself for three years, and all she needs is tissue to do the experiments to work on an unusual lymphoma-related condition called Langerhans cell histiocytosis. There's plenty of the tissue around, but she's got to get ethical permission from all sorts of places, and she's pissing about, writing these ethical things instead of doing the bloody work! It does make me very annoyed, this sort of thing.

Most of these rules came in post-Alder Hey, but how much damage did the original Alder Hey controversy do to you?

To me, none at all. To the profession, quite a lot. And to paediatric pathology, a great deal. I know of people who have given up being paediatric pathologists because of this – because they couldn't stand the kinds of pressures they were under: the phone calls, the abuse they were getting as they walked home, and this sort of thing. It has really kicked paediatric pathology in the teeth. Here it was Frank Carey who had to deal, on behalf of the Trust, with irate relatives. Frank is a really good guy, easygoing and a first-rate diagnostician, and it took a great deal out of him. He had to meet with families – not a lot of families, because it was small scale here compared with many other places. Some were perfectly reasonable, but a few wouldn't listen;

they ranted and raved, and he had to sit there and be polite and reasonable in the evenings when he should have been back home with his family. It got me quite annoyed. But personally, apart from making research more difficult, and making it more difficult encouraging people to *do* research, it really hasn't affected me directly.

Okay, to more personal things – who, apart from your father, are your heroes?

Well, I've mentioned Basil Morson to you. Alfred Stansfeld, the pioneering lymphoma pathologist, was another of my professional heroes. Alfred was the archetypal English gentleman; he would apologise to you about everything – even coming into your room! But he was hugely respected in Barts, and he's the only person I think I've ever met who, when telling you he'd made a mistake or got something badly wrong – which was very rare, I can tell you – would go up in everyone's estimation. You could palpably feel people thinking, 'Well, if *you* got it wrong, *nobody* would have got it right!' And he was very good to me when I first went to Barts, just making me welcome, giving me space, and teaching me without being intrusive.

Thinking of the cases you've seen over the years, what do you remember as being specially exciting?

Well, I can think of the amusing ones, the ones that tickled me.

Yes, give me a tickling one.

Okay. This is another serendipitous thing I got into in Barts, and what I eventually did my MD on: micro-analysis in histopathology. I used to get frustrated looking at sections because we could often see bits of foreign particulate matter in the tissues, and there's only a limited number you can stain for. You can stain for iron, copper, glycogen, but there are millions of other substances that you can't stain for. I got interested because there was a chap running the electron microscopy unit at Barts called Peter Crocker, who'd done some work on crystals in joints – gout, and pyrophosphate arthropathy and other types of joint disorders. Being frustrated seeing these particles, I'd given Peter some paraffin sections to look at; we'd started doing some basic analytical work, and begun to find bits and pieces of metal and other things that shouldn't be there. And we'd started writing papers.

One day a guy came from the biochemistry department and said, 'David, we've been sent this stone. It's been passed from this guy's bladder. We've tried to analyse it and it won't do anything. It's not triple phosphate [which is what bladder stones are usually made of]. Can you help us?' I said, 'We'll have a go.' So we stuck it in our machine and it said: silicon. *Silicon* in the bladder? I thought, 'That's ridiculous,' and I said to Peter, 'I don't believe it: the machine's playing up again!' Peter had more faith in the machine and said, 'I'll do an infrared analysis as well.' This told us it was silicon dioxide too. So I thought, 'This is crazy; bladder stones made of silicon? I've never heard of that.'

What is silicon? Where would you find it normally?

In rocks! 'But,' I said – and being a pathologist probably helped here – 'the only medicine I know with silicon in it is magnesium trisilicate, which you can buy over the counter as a remedy for indigestion.' And I thought, 'I wonder if this guy has ever taken magnesium trisilicate?' I phoned the ward, and, of course, the chap had gone home ages ago, so I got the notes out and read through. The houseman had recorded 'suffers from indigestion', so I thought, 'We might be on to something here.' I rang the patient's GP and said, 'Sorry to bother you, but do you think this chap might take anything for indigestion?' He answered, 'Young man, I have been this man's GP for the last 35 years; he has never complained to me of indigestion.' So I was very down; I was just about to put the phone down, when I said, 'Would you mind if I wrote to him, just to ask?' The GP said, 'Of course I wouldn't mind.' So I wrote, and by return I got this letter from the patient that said something like: 'Since 26 February 1945, I have taken a tablespoonful of magnesium trisilicate after every meal.' [*laughs heartily*] I was able to work out how many kilograms he'd taken in and estimate how much he would have passed out in his urine. So that was another original paper in *The Lancet* on a new type of urinary tract stone.

Again this was just about being thorough, writing and asking the patient and not being put off by the GP who'd said, 'Young man . . .' If he hadn't said, 'Young man,' I'd probably have given up anyway!

And did you write back to the GP and tell him?

Oh yes. [*we both laugh*]
We had another nice case too. Do you know what a 'bezoar' is? It's a lump of foreign material removed from the stomach – often from women who chew long hair . . .

Oh, like a cat that gets a fur ball?

Yes, exactly. People who chew string sometimes get them. Anyway, we got this bezoar removed from a young man's stomach. He admitted that he used to chew the sleeve of his duffle coat. I still had my old university duffle coat so I brought it in and we took fibres from it and tried to match them up to the bezoar fibres in the scanning microscope. Nothing matched. Then Peter Crocker said, 'You know, David, that looks to me like a coconut mat.' So we got a floor mat, took fibres out and matched them up to the fibres in the specimen . . . Exactly the same! We couldn't do the usual inorganic analysis because this was obviously organic material, but we did infrared analysis and we got exactly the same spectrum on the two samples. So we'd proven that this bezoar was, in fact, coconut matting. We told the surgeon, and he went back to the patient and said, 'Look, we think it's not your duffle coat, it's coconut matting.' And the patient said, 'Well . . . I used to chew coconut mats as well.' [*we laugh*]

So people can't hide their funny habits from you lot!

There's another story I'd like to tell and that's related to post-mortems. We always used to teach on the post-mortems at lunch time at Barts, and this story is about a patient who'd come into hospital to be investigated, and they couldn't find anything; he'd just got sicker and sicker and died.

I was supervising the post-mortem and presenting it, and I didn't know what the cause of death was – the organs were just a bit swollen, but there was nothing definite to find. I said, 'To be perfectly honest, I don't know what the cause of death was here.' And Michael Besser, the medical professor – a fantastic guy who ran a terrific endocrine unit – stood there and said, 'That's no help at all, David. That's absolutely useless; I don't know why we bothered to get a post-mortem.' I said, 'Wait a minute, Michael; there's still the microscopy to come.' I remember putting the first slide under the microscope – it was a slide of brain. I just looked at it, and every single blood vessel was stuffed with malignant cells. Then I put the lung section under the microscope, and every single capillary was stuffed with malignant cells – there was no tumour, just malignant cells in the blood vessels. I put the kidney under – every single capillary stuffed with malignant cells. It was very easy to see why this patient had died: all his capillaries had been blocked up by these damned tumour cells.

This was a very rare form of lymphoma called angiotropic lymphoma. Well, that's what it's called now, but it was still unknown then. At that time, we didn't know whether it was a tumour of the blood vessel endothelial cells, the lining of the blood vessel, or not. But we were able to do immunohistochemistry and molecular studies as well on this patient's tissues, and we were able to prove that these malignant cells were B-lymphocytes, white blood cells. So we wrote a paper on angiotropic lymphoma, which was the first time it had been described with immunohistochemical and molecular proof that it was a lymphoid malignancy without any tumour mass anywhere, just circulating around in the blood.

As a general rule, how did you feel about doing post-mortems?

Um, it took me about two years of doing them till I stopped feeling squeamish with the first cut. But always after the first cut it became a technical exercise – trying to find what had been going on. I don't think I could ever say I enjoyed doing post-mortems, but I learnt a huge amount from them, and we got an awful lot of useful information. So that made it, if not enjoyable, very satisfying to do them.

What was the significance of the first cut?

You may think it's funny but I don't like blood! [*laughs*]

Now he says!

I know; I don't know why I went into medicine!

I think, until you've made the first cut, you think of the body as a person. Once you've made the first cut, as I say, one's focus is on finding out what went on. It's usually quite easy to stay detached from the person because we've never seen them alive, and I think that's useful. I wouldn't say it makes me or others who do this any less respectful of the person, but if you keep thinking of the body as a person, then it will inhibit your ability to be clear minded about what you're supposed to be doing.

You've talked of some of the high moments in your life and work, but what about low moments?

[*Pause*] I suppose the worst moment *ever* in my life was when my first wife Jill died. That was three years after we'd moved to London, and she died suddenly and without warning from a subarachnoid haemorrhage. I went off to work in the morning and she was absolutely fine, then I got a phone call late in the afternoon to say she'd been admitted to hospital. I had to ring the hospital and was told she was dead. She must have been 37. That was certainly the worst moment in my life.

How old were your children?

Simon, the oldest, was 13 at the time. Scott would have been about eight, and the other two were in between.

How did you look after them? How did you pull your life together at that stage?

Well, I was very lucky. Five ladies who lived near us – we knew some of them through the church, and some were just neighbours we knew across the fence – they organised themselves into a rota and produced an evening meal for the kids for four or five months, which was fantastic. Then my dad and mum – my father was just coming up to retirement – came south and moved in with us. That enabled me to keep functioning. And then I was lucky enough to meet and marry Rosie about three years later.

Rosie was fantastic with the four older children, and the nicest thing that anyone has ever said to me was said by my oldest boy, Simon, after I'd been married to her for about two years. He told me, 'Dad, you choose your women well.'

So I've told you the worst thing that ever happened to me, and the nicest thing anyone's ever said . . . But bringing it back to pathology, I felt tremendously reassured by the post-mortem that was done on Jill to be told what the cause of death was.

It wasn't clear at the beginning?

No. I mean, she was well when I left, and then she was dead. I think – well, I *know* – the hospital assumed she'd taken an overdose. She was the last person in the world who would ever have taken an overdose, and it annoyed me that people could even think that about her.

And it must have hurt, too, at the time.

It did. But the post-mortem told me unequivocally that it was a subarachnoid haemorrhage. There was no doubt at all, and it was tremendously helpful to me to know that. So post-mortems can be very, very helpful to individuals – to the *living*.

With all your personal and professional experience of death, do you fear it yourself?

Do I fear death? No, I don't think I fear it; but I want to do a lot more of enjoying myself first. I know everyone says they want to travel, but there are places in this world I'm very keen to see. And I've enjoyed wherever I have been so far.

This interview suggests that, generally speaking, you get a lot of fun out of life.

Yes, I do. I do.

EASING THE PAIN OF LOSS

Irene Scheimberg
*Consultant Paediatric and Perinatal Pathologist, Barts and
The London NHS Trust*

Irene Scheimberg's great-grandfather was Russian. There was discrimination against Jews in education and so, determined that his children should get the chance to go to college, he emigrated with his family to Argentina in the early 1900s. Scheimberg grew up in a highly politicised, intellectual home, shared with her grandparents. Midway through her medical training she had to flee Argentina as the military regime began arresting and killing close friends. She found work to support herself through the final years of medical school in Spain, and then came to the UK, where, self-conscious about her English, she chose to specialise in pathology, partly because she reckoned it would involve less direct contact with patients. She was offered a post in paediatric pathology at Great Ormond Street Children's Hospital and then began doing perinatal pathology – studying the diseases and causes of death among fetuses and newborns – which, 20 years ago, had precious few practitioners.

Scheimberg's experiences in Argentina are a reference point in everything she does, and her priority is to help families understand the reasons for a child's illness or death and to cope with their loss. 'I do believe that no matter how much science benefits from what I do in trying to find the causes of disease and malformations, the most important thing for the people who are at the other end of what I do is the emotional support that my results can give them,' she says.

——•——

My grandfather was one of the founders of the Argentinian communist party – but he left within two years because he couldn't stand the fact that you were programmed to a particular way of thinking. We think about communism the way we do now because we *know* what it did to the Soviet Union, but we're talking about 1919, when the idea was that you were there to help other people, that everybody was going to be equal . . . It was a wonderful dream that turned into a nightmare, so he became a civil rights lawyer, defending freedom of speech and things like that. I always wanted to do something that could help other people. It was a family in which you didn't do things just for yourself, you had a bigger vision – you *had* to do something for society.

So did the politics really wash off on you?

It was part of everyday life.

What made you decide to do medicine?

My parents are both doctors. I always liked science, biology. I like to understand how things work and why things happen. Everything has a logic and a reason . . . In a way, being a doctor gives you some protection against your fears of disease, because at least you know the mechanism of what goes on. And my parents liked what they did, although they were in completely different branches of medicine. My mother is a child psychiatrist, and my father is a gastroenterologist.

But my other great love was history. I was going to put my name down for both at university, and then I thought: do I really want to study history in a country where the next dictatorship is going to tell me which books I can and cannot read? No, I don't. I can study history on my own. I decided I was going to do medicine, and I have never regretted it. But it's a lot of responsibility – and sometimes it can be very burdensome, because you know that a child's life depends on your diagnosis.

How much are you aware of the human being behind the image under the microscope?

You're always aware that there is a child at the other side of a slide. When I chose medicine, it wasn't because I thought it was interesting to see pink things under the microscope. I studied medicine because I thought that there was something I could do to help other people – so they are never away from my mind. I *know* there is a person, and in my particular case it's a child, on the other side of that slide, whether I meet the child or the families or not. But I like seeing the patient – I like not to lose this human contact, because for me that's what medicine is all about.

Tell me a bit about your departure from Argentina?

Oh dear, I'm going to cry, I know that! I'm sorry, but you asked. In 1976 there was a coup d'état, and a military dictatorship took over that was the most brutal in Argentina's long and brutal history of military dictatorships. Lots of my friends were 'disappearing' off the streets, or being obviously killed, and my ex-boyfriend Carlos – a very recent ex, and somebody whom I really loved – disappeared one day. I kept phoning his mum until one day she told me that his body had been found in the river, with marks around the wrists and ankles, and he was being buried. I went to bury him, and 10 days later I was on a plane to Spain. The police or the army, or whatever, had been to check on his friends, and, of course, my parents were very concerned. So was I, and so I had to leave. Lots of my friends were killed or 'disappeared', but the death of Carlos, that was the particular reason why I left.

I went to Spain and I had to find myself a job. I was 21. I came from a very close-knit extended family, where you always had somebody to rely on, and suddenly I was on my own. The first winter there, I have this image of myself sitting in the tube, it was January, and I just wanted to cry – it was all so depressing. Then my aunt came to Spain with news that my best friend, Silvia, had disappeared. When I left she was pregnant, and now nobody knew where she was. I cried and cried, and it was awful . . . There's a book that was published a few years ago about the students from my school who disappeared, and there are 105 names in it. I personally knew 35 of them, and some were very good friends.

When I left Argentina, I had done two years of medical school. I managed to go back into medical school after a year in Spain without losing too much, and I eventually qualified in Spain.

So what has that done to you – living through a period when your own friends were disappearing? How has that affected your philosophy of living and the work that you do?

I don't know. I think it might be two contradictory things: on the one hand, I think I do have personal understanding of what parents go through when they lose a child. At a bereavement conference some time ago they read out something that a mother said on the death of her baby. She said that not only did her family lose a four-and-a-half-month-old, they lost the toddler, and the child who would start school, the pimply adolescent, and the wedding and the children and the grandchildren . . . [*she weeps*] I can identify with that because sometimes I think about Carlos' children that could have been mine – or somebody else's, it doesn't matter – but it's the future. And it's the loss of the future that is so painful. So on one hand, my experience makes me understand what loss is. On the other hand, it makes me very intolerant of people who complain about minor things. It makes me just want to say, 'Come on, get on with life.'

Tell me, how did you finally end up in London, not Spain, and practising paediatric pathology?

Well, I met my first husband, Luis, in Spain. Luis was a very nice guy. He had a job at a Spanish bank, and because he spoke English they decided to send him to London. I had finished medical school, so I came too. That was 1984. Spain was not part of the European Union, so I couldn't work as a doctor here. But my dad said, 'Why don't you talk to Julia Polak?' Julia's uncle, Moisés Polak, was a famous Argentinian neuropathologist who had been a friend of my father. And my father had been Julia's tutor at university. So that's how I went to work with Julia. I was given the job of a research assistant, and I learnt to do *in situ* hybridisation. I worked there for six months, and I met really nice people; some of them are still friends.

We went back to Spain, and then Spain got into the EU in 1986. Luis moved heaven and earth to come back to London, because there was nowhere else like it for him. So we returned, and mine was one of the first Spanish medical diplomas to be recognised by the General Medical Council here.

I had a Spanish passport by then. When I had to go and swear before the Constitution – and the first time I voted – I was so proud! I've been in England now for 21 years and I *love* this country, I really do. Sometimes I think: should I get a British passport, because I should be voting here? But I keep my Spanish passport, because of a sense of loyalty to the country that helped me when I really needed it most.

So I got my qualification recognised, and when I knew that we were coming here for a long time, I got some letters of introduction to people. I got one to Salvador Moncada, who discovered the role of nitric oxide in the body. I got a letter to a person who was doing research into neuropathology. And I got a letter to this fantastic Spanish pathologist called Fernando Paradinas, who looked like Don Quixote – the idea of the Spanish gentleman, thin with a long face and a beard. He was one of the best pathologists in the UK.

The neuropathologist didn't pay any attention to me. Salvador didn't offer me a job, though we later became very good friends. But Fernando Paradinas, who worked at Charing Cross Hospital, said, 'Come and see me.' The work I'd been doing with Julia Polak was extremely useful – they offered me a job as a research assistant with Bernard Fox at Charing Cross, and said that in the meantime I could go and learn some pathology. Fernando used to be called 'The Spanish Inquisition', I remember! He used to organise the 'black box'. This happens in all pathology departments in the UK: basically the black box is where you put the week's interesting cases, and then you give them to the junior doctors with a minimal history, and they have to make a diagnosis. The cases are then discussed by the group once a week. The system trains them to diagnose.

I was the only female among the junior doctors at that time, and I remember Fernando saying with a smile, 'Let's see if we are wasting our time teaching Dr Scheimberg.' He put a slide up on the television, and it was the appendix of a 35-year-old woman. I had seen very few things by then, but

I thought, 'He wouldn't be asking me something he doesn't expect me to know. The only thing I've been looking at so far is female genital tract pathology. It must be endometriosis.' So I said, 'It's endometriosis,' and he said, 'Oh, so we're not wasting our time!' It was just luck and reasoning, really.

Fernando was a fantastic teacher. He knew *everything*, but he was incredibly modest. Everybody who went through Charing Cross will tell you about him. We were all a bit in awe of him. You'd show him something and he'd say, 'Mm, I saw this in 1965 . . .' He had this fantastic memory. He had a collection of slides from 20 years before that we all used when practising for our exams.

And did you think: this is the direction I want to go?

Well, pathology was something I always liked, because I thought: if you go for a speciality, you'll learn everything there is to know about the heart, say, or the kidney or whatever, and forget about the rest, whereas in pathology you keep a grasp on everything. But it's interesting, because now pathology also is divided into sub-specialities, and you have people who *only* do gastrointestinal pathology, or whatever. Maybe that's why I like paediatric pathology, because, again, you have to know a bit of everything.

I did my training in paediatric pathology at Great Ormond Street Children's Hospital, and then when I came here to Barts and the London Hospital, I started also doing perinatal pathology – which is a *completely different* thing. Basically, in paediatric pathology you deal mainly with living children, and you try to help to keep them healthy and alive. Perinatal pathology is about death and loss. And . . . it's *different*.

How exactly does paediatric pathology differ from adult pathology, and from perinatal pathology?

Children are not dwarf adults, okay? They're not little adults, and they've got different kinds of diseases. It's like a parallel world: you need to focus on different diseases and different things in paediatric and adult cases.

Then perinatal pathology is a completely different kettle of fish again. The amount of obstetrics, neonatology and physiology you have to know is quite substantial. I always say to the students, 'When you do a perinatal post-mortem, it's as if you are trying to make up a puzzle in which lots of little pieces are missing. Every piece of information you have, you'll need in order to build a picture and get an idea of what happened.'

Is perinatal pathology fairly new as a specialisation?

Well no, but for a long time there were very few specialists – people like Jonathan Wigglesworth or Ian Rushton or Jean Keeling – and in most hospitals the babies were left to juniors or whoever was available. Then about 20 years ago a group of people started to get interested in perinatal cases. They realised that it wasn't just a 'left-over', it was very interesting, and it was something that could help people who were absolutely devastated by what had happened.

So before then, what happened to babies who died in the uterus or just after birth?

If mothers were in the right hospital, they'd get a very good report from an eminent person like Jonathan Wigglesworth. But what would have happened in most other hospitals 30 years ago is that someone would have done a post-mortem and said, 'Well, there's no malformation; I cannot see any major infection.' And that would be it; they couldn't say any more about what killed the baby. But nowadays there are very, very few cases in which we cannot say something useful for the next pregnancy.

For instance, I had a case in which a woman had three early miscarriages. I get the fourth miscarriage, and I look at the history and think, 'God, this is the fourth miscarriage in four years, all before 20 weeks.' You look at it; you see the typical pathology and you tell the clinician, 'Be careful, this woman might have lupus [an autoimmune disease].' Now if you're not a specialist you will miss it.

So it's a special skill. Nowadays, I cover most of north-east London and some areas around it. So we take care of all these babies, and you can tell the difference in the level of satisfaction, because when you start doing this work for a hospital, they'll say, 'Oh, we'll only send you about 20 babies a year.' And after two or three years, when they're getting results – and the clinicians can show results to the parents so they're much more confident about asking for consent – you go from 20 to 40. We've got a hospital in a very high-risk area which sends us up to 60 babies a year.

In my own hospital, when I started 11 years ago there were eight post-mortems a year, and now there are at least 25.

What would you say are the principal purposes of a post-mortem?

Well, the main reason to do a post-mortem is for the family. And that's something you should never lose sight of: you want to tell a family what happened. I think when a child dies the parents tend to feel guilty. As an adult, a parent, you're supposed to be able to protect your child, and the child's death is like a failure to protect. So it's very important that there is a professional there to tell parents that this would have happened, no matter what they did, because that guilt is a terrible burden for a parent.

The post-mortem is the first and only time that that child – particularly if it's a very young baby – will be examined by a doctor. We *are* that baby's doctors; we have a duty to them and to their families, because what we find will not only help the parents in their grief, it might also help them if there's the possibility of the same problem occurring again in another child. So a perinatal post-mortem is not just a closure; it's a closure and a continuation, because it helps families to carry on with their lives, and have more children.

Can you give me an example?

There are several cases, though they may not sound so dramatic. I did a post-mortem on a 10-week-old baby, who had a very rare heart condition. Had the post-mortem been done by a non-specialist, this would probably have been missed. But it meant the family could have proper counselling about the chance of the same problem occurring again. And I've got a photo on my wall of a woman who had an autoimmune disease and lost her baby. What we found at the post-mortem allowed the obstetricians to rescue the second baby, who had to be delivered prematurely but was fine. I've got a photo of that second baby aged two, and these are really very rewarding moments.

So how often do you think that the cause of death on the death certificate has been inaccurate?

The Confidential Enquiry into Stillbirths and Deaths in Infancy found that '. . . in 70% of post-mortems on children done by a non-paediatric pathologist, essential tests were omitted, and the diagnosis was deemed to be incorrect in approximately 20% of cases. And these led to failure to recognise inherited conditions and on occasions led to inappropriate suspicion of harm.' So the post-mortem is very important for audit purposes also.

As I said, the most important purpose is for families. But post-mortems are an important element of teaching and training too. Medicine doesn't only progress on big, wonderful discoveries that may bring people the Nobel Prize. Medicine progresses little by little; it's like building a wall in which every little piece of knowledge is a new brick – and pathology can give a lot of bricks to that wall, as well as helping families.

Let me give you an example: 20 or 30 years ago, children were being operated on for congenital heart disease and the death rate was quite high. Now the death rate for the same malformation has come down dramatically. And the reason is that the hearts of the children who died after an operation were being kept and they were being studied by the paediatric surgeons who had done the operations. Little by little, they were discovering the abnormalities in these hearts that might not have been so obvious at first. They were realising that maybe they should put the stitches in this place and not that. And maybe the conduction system was also in the wrong place. And as a consequence of those investigations, the mortality in congenital heart disease has decreased dramatically.

So, as I say, what we discover is not very dramatic: it's not the kind of thing that makes the headlines or gains the Nobel Prize. We're the workers, like the ants. We're just adding the little brick that might not look important in itself, but when you see the whole wall, you realise how important it is for that brick to be where it is. I mean, what we do is not at the cutting edge of research, but it is what cutting-edge research is based on.

Are you yourself working on any particular little bricks at the moment?

Well, there's this big controversy about shaken baby syndrome, and one of the things I'm trying to find out is the mechanism of bleeding in the brain in babies that do not have trauma, and whether hypoxia [insufficient oxygen to the brain] has a role to play. What we see is that in lots of cases of babies who die in the uterus or during the birth process – circumstances in which there is no particular trauma or violence – there is bleeding, particularly in the dura, the membrane that covers the brain.

I want to understand *why*, in cases of hypoxia, you get haemorrhages. Is it the same mechanism as in abuse? Or is it a different mechanism? We don't know. At the moment we are seeing if we can take a little of the vitriol out of the controversy, and try to behave like objective scientists. If we find out the mechanism, we might be able to distinguish some cases from others.

You see, the problem with the shaken baby controversy is that it's very dogmatic. If I don't accept religious dogma, and I don't, I'm not going to accept scientific dogma. If it's there, it can be proven. I need to understand the mechanism. Because although I *do* recognise that some parents are capable of doing very nasty things to their children, I'm very uneasy about people just saying, 'Oh, if it's got subdural haemorrhage, retinal haemorrhage and brain swelling, it can only be shaken baby syndrome.'

Have you got clues as to what the mechanism might be?

I've got all sorts of theories that I need to explore. We know that in young babies the capillaries are smaller, more immature. We know that hypoxia also changes the behaviour of the endothelium and the blood vessel wall; we need to know to what extent. We always say that children are not small adults; we need to understand the different cut-off points in the development of a child that make things different. For instance, a baby inside the uterus won't have the same mechanisms as a newborn baby, although there is a transition. And a newborn will be different from a three-year-old, so lumping together all children under three, for instance, doesn't help.

I think, in order to understand the different mechanisms in the production of similar things, you have to understand the environment in which they occur. I have a talk that I give in which I pin up a map of Rome; and you know the saying, 'All roads lead to Rome'? I say instead, 'More than one road leads to Rome.' And that's what we have to explore: the roads.

Do you find that getting consent to keep autopsy material for research and teaching is particularly difficult since Alder Hey?

Well, maybe because of the kind of family I grew up in, I am a strong supporter of people's rights. But rights come with responsibilities; you cannot divorce the two. People are part of society; they benefit from what other people are putting into that society, so they have a duty to give, especially if it doesn't imply giving anything extra. When I ask for consent I tell people, 'The blocks and slides that I have taken for the diagnosis to try to help you and your family can be discarded after diagnosis. Or they can be used to teach other people that will continue my work.' When they say yes, they feel part of the society and they are fulfilling their duty.

When people say no, I think it can be for several reasons. One is that when they are asked for consent, things are not explained to them properly. Another reason is that they are really hurt; some people are very angry and feel the system has failed them. Another reason is superstition: 'God knows what you're going to do with this material . . .'

I'm really unhappy that when the revised Human Tissue Act came into being in 2004, they did not treat the small blocks of human tissue that we take at post-mortem the same as tissues taken from living patients. I'm not talking about organs: I can understand how people feel about whole organs, although I don't agree. My feeling is that when I die, my organs have two possibilities: either they rot and they're eaten by the maggots and go back into the cycle of life, or I'm cremated and my ashes will eventually go back to earth and back into the cycle. And I'd rather have somebody learn something from them than be eaten by the maggots!

As I told you, I'm not a religious person, but if I *were* and believed that the body is the vessel that carries the soul, I'd argue that the soul doesn't need a piece of liver or what have you to reconstitute itself. Otherwise what would happen to people who suffer amputations in war, or lose an eye or something? What does that mean? That they won't be whole when they go to Paradise? I was disappointed at the way religious leaders were reluctant to confront this upsurge of superstition in people. They could have explained that, yes, it was wrong that things were not explained; it was wrong that the proper consent wasn't taken; or that not enough attention was paid to it. But they should have assuaged people's fears about the effect on their souls of losing a little piece of liver or spleen or whatever.

We have to separate things. Not requesting proper consent and not explaining things to people was wrong, and I think there have been big improvements. But I don't think that even the parents of Alder Hey had envisaged that you'd have to ask consent for teaching and training on blocks and slides that are taken anyway for diagnosis. We could end up without having anything – not being able to teach anybody, and nobody to do our job in a few years' time. I hope that at some stage it will be recognised that it's impossible to carry on without imparting knowledge to the future generations.

What's the purpose of keeping tissue such a long time? What's the value of historical samples?

There's a recent example that relates to patients who have problems of constipation in childhood for which nobody can find a cause. They go on to be adults with very serious problems, because it's not just the constipation, they get sudden diarrhoea as well. And people do biopsies, and they keep telling them that there is nothing wrong with them; that there must be something wrong in their mind. And there are even people who have committed suicide because of this. Now one of my colleagues, who is a professor of pathology here, Jo Martin, recently discovered that there is an abnormality in the muscle wall of the intestines of these patients. Because we have tissues from years ago she has been able to go back to patients who had biopsies 15 years ago – and for 15 years have been told that they were mad – and she's been able to prove that they were not mad, that they had an abnormality that 15 years ago we could not diagnose. Now you tell me, how important is it for these people to finally have a diagnosis? One day we might find a cure, but just to be able to tell them, 'No, you are not mad, you really had something wrong with you.'

On a more personal note, your own experience of losing loved ones has obviously given you a great deal of empathy with families in your work. But how were you able to put the traumas of your time in Argentina behind you?

I had been extremely successful in putting all my dead friends into a drawer, closing the drawer, locking it and throwing away the key, because I had to survive. I was 21 years old; I couldn't carry my dead with me. So I never properly grieved for my boyfriend, for all my friends, for losing my family, my country – my life. I never grieved, I just carried on. Alder Hey had started to move a lot of things in me, and then the Twin Towers collapsed in 2001 – and I was obsessed. It was as if all my dead people came shouting back at me.

I had all this grief coming back to me with the collapse of the Twin Towers; I had to deal with it, and I was very stressed. During that period, I remember doing a post-mortem on a child who had died of meningitis, and crying as I went to casualty to get my prophylactic antibiotics – I couldn't stop myself from identifying with the family. It was the worst post-mortem . . . He was a child about the same age as my own Pablo, eight years old at that time; he had the same build, the same hair colour.

Then, in 2003, I had 11 baby post-mortems in one week; I was very stressed, and I had problems with some junior doctors. I just couldn't cope any more; I had to take three months off. My present husband Charlie was very supportive and helped me get better.

Generally speaking, can you find the emotional distance from your patients?

I think it will always be difficult if it is someone close to my son Pablo's age, and who sort of looks like him. I did a case of a very sick child who was 11 years old, but he had been ill all his life and really looked much smaller. It was awful for the family, but for the child, dying was probably a kind of liberation and release. I don't have difficulty in those sorts of circumstances. But it's these children who are completely okay and then something happens to them – it's devastating. But you have to find a way. If you are not stressed, you can absorb the pain. But when you are stressed to the limit that can be the last straw.

It was interesting when I went back to work. As a paediatric and perinatal pathologist, you are very isolated and people don't realise what work you do, but those three months that I was away, they realised that they couldn't cope without me.

Tell me, how much of your personal identity is tied up with being a pathologist?

I am not a pathologist; I am a woman who works as a pathologist. I am not defined by my job. My job is part of who I am. Actually, perinatal pathology is quite a big part of who I am, because it allows me to do my bit for people while I do my work. I always think that I'm helping people at probably the worst time of their lives.

And that's the main motivation for what you do?

Yes, it helps a lot. There are lots of intellectual challenges in medicine, and you can learn to like whatever you do, if you put enough into it. I think the extra thing about helping people at a time when they lose what is most precious to them has to do with my past, with my losses when I was in my twenties. Somehow I can identify with what people are going through, and that's important to me.

At one point during the Alder Hey crisis I said, 'I am going to go and talk to the Liverpool parents, so that they realise that not all pathologists have horns and are horrible.' At the beginning they were all very confrontational – there were lots of them – and I said, 'I do understand what it is to experience the untimely death of people.' And I told them my story – because they were so immersed in their grief that they didn't realise other people might have had a traumatic history as well. They were surprised because I was crying. One of them came up and hugged me afterwards, and said, 'I never thought I'd ever hug a pathologist.'

STEM CELLS AND THE BODY'S REMARKABLE CAPACITY FOR REPAIR

Nicholas Wright
*Warden of Barts and the London, Queen Mary's School of
Medicine and Dentistry*

Nicholas Wright is unusual in that he knew even before he went to medical school that he wanted to be a pathologist. He traces his career choice to a book he read as a teenager – *Men Against Death*, by Paul de Kruif – about the early bacteriologists, Louis Pasteur, Robert Koch and others, who fired his imagination. 'That golden era of bacteriology from . . . 1880 till 1920, when most of the bacteria that cause diseases were discovered, was extremely interesting,' he says. 'It's a real saga and I think any young person . . . would be turned on by it.'

On further reading, Wright realised he was more interested in the effect the bacteria were having on the tissues than in the bacteria themselves, and so he chose to specialise in histopathology. Besides running a department and nurturing new generations of pathologists, his passion since the early 1980s has been for stem cell research. 'I was really interested in how tissues were put together and how they responded to insult and did repair . . . That led naturally to stem cells,' he says. Wright and his group are at the cutting edge: they were the first to discover adult stem cells, in the bone marrow of humans, that can transform themselves into a wide range of tissues when the body needs them for repair. Until then, most scientists believed embryonic stem cells alone possessed such 'plasticity', and that adult stem cells were specific to each tissue type. The discovery, published in *Nature* in 2000, unleashed 'an avalanche of thought', because of its potential to provide novel treatments for disease. But for this to become reality, he says, 'The trick is to find out what triggers [the process] and try to expand it.'

Professor Wright, can you start by giving me something of the history of pathology as a specialisation – a historical perspective?

I think that before the Second World War, most hospitals had what was called a general pathologist. And that general pathologist would perform post-mortems, would do what's called clinical biochemistry, examining blood and urine, and haematology, looking at the blood, and would also look at micro-organisms – bacteria – and isolate and diagnose infectious diseases. So they were very much generalists.

After 1945 it became quite clear that nobody could do everything, so we had specialisations. Very rapidly we got the evolution of what are now called histopathologists – they're the people who perform autopsies, analyse tissue sections, look at breast biopsies, look at things like cervical cytology. And other people spun off into being specialist haematologists, microbiologists, clinical biochemists, virologists. So now we have a cadre of people who specialise in each of these disciplines and give a very highly competent service, without which the NHS just could not function.

Going back maybe a century, were ordinary doctors trained to do these things as well, or was there still some special person who worked behind the scenes helping with diagnosis?

I think 100 years ago certainly there were pathologists, but again they were generalists in the main. Pathology as a discipline evolved with individuals like Rudolf Virchow in the 1840s, who really established the basis of pathology and disease. Virchow was a polymath: he started off as a medical practitioner, then became a pathologist. He was the first person to recognise thrombosis, for example, and the fact that pieces of thrombus could break off and become what are called emboli. He analysed the cellular basis of pathology and really introduced the microscope into diagnosis. He later became an anthropologist and a politician. However, even the best pathologists can make mistakes. The Kaiser had a lesion in his larynx, which Virchow diagnosed as a benign condition. In fact, the poor Kaiser had cancer of the larynx, and he died!

You've been called 'doctors of death' by the general public, who think of you mostly in connection with post-mortems. But what proportion of your work is actually spent dealing with dead bodies and what proportion in dealing with specimens from people who are alive?

I think most pathologists these days do very few autopsies. For many, many reasons, the incidence of hospital autopsies – the post-mortems done on patients who die in hospital – has been declining: methods of diagnosis have improved; there's been a natural tendency not to ask for autopsies; and certainly after the Alder Hey disaster – or debacle, whichever way you look at it – the willingness of hospitals to ask for them, and also the willingness of relatives to give permission, have lessened. So there are very few autopsies done at the moment. It's very worrisome because it's a major teaching vehicle for both medical students and junior pathologists. But to answer your question, the thing that most people like myself, who are histopathologists, do will be to diagnose diseases in living people.

Let me give you a graphic example. One of the worst things that can happen to you is to have a malignant condition of a bone – an osteosarcoma. Now these often present in young people who may have had a knock, or found a lump sticking out of their lower leg or something. They go to the GP and he or she will examine it. The GP doesn't like the look of it and will send the patient for an X-ray. Sometimes it has a characteristic appearance, sometimes it doesn't. But what will happen is that the young person will be admitted to hospital and, usually under general anaesthetic, a biopsy will be taken of the bump. Interpretation of the appearances of that lesion is often not straightforward; it's a very specialised and very difficult job. And, of course, a great deal depends upon it, because if it is an osteosarcoma, although there are now modern techniques for preserving the limb, it usually means an amputation, which is a traumatic and serious operation for a young person. The decision about whether that lesion is benign or malignant is purely the pathologist's responsibility. So he or she looks down the microscope and says, 'Yes, this is malignant,' and once that diagnosis has been made the die is cast. There is no way back. You can't say a few days later, 'I'm sorry, I made a mistake.' So the pathologist plays a very important role in determining exactly what happens to *you* as an individual. The histopathology service is really central to everything that surgeons and physicians do.

Tell me, what sort of family did you grow up in? Did you follow in the footsteps of other medics?

No, there wasn't a single medic in my family. In fact, no one to my knowledge had ever been to university. My father was by profession a sheet-metal worker. He was badly injured in the war because he was working on Lancasters, so we had a bit of a hard childhood. I won a scholarship to Bristol Grammar School from, I suppose, quite a poor home really. We were on supplementary benefit and all the rest of it. My father was in hospital for a long time – three or four years – and then he came out and he went back to work again. But he couldn't really work very well, and he was an invalid most of his life after that.

How many of you were there?

There was myself and my sister. My sister became an actress, and I went off to medical school. My mother came from just an ordinary Bristol family. Her father worked on the boats down at Avonmouth, unloading enormous bags of corn and stuff, so we were just a very ordinary family.

Did your parents believe in education, and were you encouraged?

Oh yes, very much so. I mean, it was always tacitly assumed I would go to university. My parents didn't know what I was going to do – I think my mother wanted me to become an architect, because I was quite good at drawing.

I entered medicine with the express wish to become a pathologist, which is very unusual. It sounds rather poetic; it's not really! But as a 15-year-old I read a book called *Men Against Death* by Paul de Kruif. This was really the history of the early bacteriologists, like Robert Koch, who became my hero. And when I went to see my careers master and said, 'I want to become a pathologist,' he said, 'Well, you have to read medicine.' When I applied to medical school, they said, 'Why d'you want to do medicine?' and I said, 'I want to be a pathologist.' They replied rather condescendingly, 'What do you know about pathology?' So I told them! I started with the express wish to become a pathologist, and I've never ever regretted it.

What was so fascinating about this book?

It was the fact that you had these conditions – like Rocky Mountain spotted fever, tuberculosis, leprosy, typhoid, cholera – which just seemed like black magic for the practitioners of those days. They had no idea what was going on. Then along came Louis Pasteur and Robert Koch and the early bacteriologists, who isolated these organisms and put medicine on a scientific basis for the first time. That golden era of bacteriology from, I suppose, 1880 till 1920, when most of the bacteria that cause diseases were discovered, was extremely interesting. For example, to go from looking at diphtheria to isolating the organism *Corynebacterium diphtheriae*, developing an antitoxin, and then an antibiotic – it's a real saga, and I think any young person reading that sort of thing would be turned on by it. It's extremely exciting.

Now the same thing is happening in genetics. We had these conditions that were a black box and now we're understanding the way they're working, and hopefully in the future we can actually correct those gene defects. It's a really romantic story if you think about it.

If it was the bacteriologists who fired your imagination, how did you get into the branch of pathology you're doing now, histopathology?

When I got to know more about it, I became more interested in the *effects* the bacteria were having on the tissues rather than the bacteria themselves. I was reading around all this before I went to medical school, then when we did pathology as undergraduates my feelings were cemented. So I just did my house jobs and went straight into pathology.

Tell me about your research – what did you start with and how did you find your path?

Well, I taught myself experimental design, basically. I wrote these books with a colleague of mine [*he reaches one off the book shelf*] – The Biology of Epithelial Cell Populations – I was really interested in how tissues were put together and how they responded to insult and did repair. So I became interested in the cell cycle and how cells divided, and that led naturally to stem cells. I suppose stem cell biology has been my interest since the early 1980s.

What was the state of knowledge at that time about stem cells? What was known and what were you interested in teasing out?

I was interested in the gut basically, the gut and the skin. There was very little known about what stem cells did – whether they were able to give rise to multiple lineages, or whether there was a stem cell for each lineage.

These are not the embryonic stem cells?

No, no, these are adult stem cells, tissue stem cells like those you find in the brain, in the skin, in the gut, in the liver . . .

Can you explain that a bit further, because I think a non-medical person will be quite surprised to hear that there are stem cells everywhere. Where are they, and what's their role?

Well, embryonic stem cells are found, obviously, in the embryo. And you have to have a fertilised ovum, which is where all the ethical problems come in. But when the human or animal is born, all the tissues contain cells that are stem cells. They're cells which are able to give rise to all the cells in that tissue. For example, the stem cells in the lungs are able to give rise to all the lung cells, and the stem cells in the gut give rise to all the gut cells, etc.

And they're there for replenishing?

Yes. Many tissues turn over very rapidly – the bone marrow, the skin, the gut are renewing very rapidly, so they have to have a reserve of cells to accomplish this, and the stem cell is the sort of 'governor'. It produces cells and it sits in a thing called a 'niche'. It lives there all the time, and every time it divides, one daughter cell goes off to produce more cells to be replenished. There are many of these niches in the skin and the gut.

What do they look like under the microscope?

Just like ordinary cells. But I mean it's only now that people are getting a handle on what they actually are – what their markers are and how they behave – and are working out ways in which you can study them in animals and man. Later it became clear that the idea that adult stem cells were fixed forever, giving rise to the same sort of cells always, was *wrong*: they have remarkable plasticity. Bone marrow stem cells can give rise to skin cells; they can give rise to liver cells; they can give rise to lung cells. My lab has done quite a lot of that work as well.

Yes, I understand your lab was one of the ones that actually discovered this plasticity. Tell me how that happened. Were you looking for it or did it happen serendipitously?

There were some early suggestions in the late 1990s that if you gave bone marrow transplants to animals, then you eventually found cells in the liver that came from the donor.

So we had the happy idea that we could find that in humans too, because humans have liver transplants, they have bone marrow transplants. So we got patients with bone marrow transplants and patients with liver transplants and we were able to identify the transplanted cells . . . We were able to follow exactly where they were going. So we did that in the liver, in the kidney, in the gut in humans . . .

But where had they come from, these cells?

They came from the bone marrow. So if someone has a bone marrow transplant – for example, you want to replace the cells in the bone marrow in someone who's suffering from aplastic anaemia [their marrow's not producing enough blood cells], so you give them a bone marrow transplant – you colonise all the marrow and you start producing blood cells again. But the cells also go elsewhere – they go to the liver, they go to the kidney and they go to the gut – in small numbers. And the bone marrow cells become kidney cells or liver cells or skin cells as well . . .

This wasn't understood until you discovered it?

No, not until about the year 2000. And we were the first to show that this occurred in humans.

And how exciting was it when you actually began to observe this?

Oh, it was very exciting! It was published in *Nature* – very exciting at the time. Then people got a bit blasé about it, called it old hat. But you know, there are more things being discovered now as a result. American groups are now claiming that when you get tumours being produced in animals – for example, in the stomach – they really come from the bone marrow. So it started an avalanche of thought.

D'you think this is true?

Well, it's a very interesting idea. *Helicobacter pylori* is the organism that causes gastric and duodenal ulcers. These investigators did a bone marrow transplant from male into female [this allowed them to track the transplanted cells because they carried the Y chromosome in their DNA]. They then infected the animal with *Helicobacter pylori*, and that gave rise to inflammation and then cancer. And they found that the cancerous cells actually came from the bone marrow that they'd transplanted, which was very strange.

The *Helicobacter* infected the stomach and damaged the animal's own stomach cells, and the cells in the bone marrow came in and took over and then produced a tumour. So it's a very bizarre thing.

When you first discovered this phenomenon you hadn't anticipated it?

No. We thought that bone marrow stem cells give rise to bone marrow; gut stem cells give rise to the gut; kidney stem cells give rise to the kidney, etc. We didn't know there was this ability to *change* tissues, basically.

And how much plasticity do you believe there is now?

Well, I think there's a lot of plasticity *potentially*, but it doesn't happen very often. The trick is to find out what triggers it and try to expand it. I mean, if you give a bone marrow transplant to an animal and you damage the skin, about 14% of all the cells in the new skin come from the bone marrow transplant. In the kidney, if you damage it, it's about 7%. So it's not a large number. But there are animal models of liver disease where almost half the liver could be taken over by the bone marrow transplant. It depends on the amount of damage you're able to produce. In the intestine, we've shown that some 60% of the myofibroblasts – the cells which support the epithelium – come from the bone marrow and that in colitis this is increased. And, we've also found whole blood vessels of bone marrow origin.

This is an attempt to repair the damage, is it? A sort of maintenance programme that goes a bit haywire?

Yes. If you want to be teleological about it, there's a reserve of cells in the bone marrow which, when things are going wrong, can actually produce stem cells for other tissues. That's the concept behind it. It's pretty controversial; not everybody accepts this. But there's a growing body of evidence suggesting that it's true.

So what do you have to do to convince people?

Well, it is widely accepted by a lot of people, but there are several leaders in the field who doubt it. Yet 30 or 40 groups have shown that what we did in the gut is true. So I think it's highly likely to be true.

How much hands-on research do you yourself do now, and how much are you managing a team of people who are doing the microscope work?

A lot of this is molecular – highly detailed microdissection and gene sequencing – and I don't do that. I just come up with the ideas. But I plan the experiments with clinical fellows, PhD students and post-docs, and help them to analyse them. And if there's any problem with the interpretation under the microscope, then I will do that, because we're looking at diseased states – gastric tumours, colonic tumours – and I'm the one who has to say, 'Well, this is an adenoma; this is a carcinoma.' So if you're a pathologist running a basic science laboratory like I am, you're in a very happy position, because the scientists depend on *you* to tell them what the tissue is. And you have the medical background to say, 'We should do this, because this is important clinically.'

One thing you haven't done so far is to give me any examples from your case book . . .

Well, I don't want to disappoint you, but I'm not that sort of pathologist! I've never been turned on by individual cases, I've been turned on by *disease*. The things that have made me think a great deal are the appearances you see with just ordinary cases – when you're looking at a duodenal biopsy, when you're looking at an ileum biopsy in Crohn's disease, you see changes and you think, 'What is going on there?' When you're looking at a Barrett's oesophagus, you think, 'Why is that how it is?' In Barrett's oesophagus, for example, and in the stomach, most of the common cancers don't actually develop straight from the ordinary epithelium. The epithelium *changes*, and that change is called a metaplasia. So the stomach epithelium changes into intestinal epithelium (it's called intestinal metaplasia), and the lining of the oesophagus becomes intestinal metaplasia too. Now, trying to work out why that happens is the thing that really turns me on. So it's not individual cases, it's *phenomena* . . . Looking at the stomach, you ask, 'Why does this gastric epithelium suddenly change into intestinal epithelium? And why is the intestinal epithelium more prone to get cancer?' Those are the questions. Many pathologists will just say, 'Right, intestinal metaplasia. Next case.' But my way of thinking is, 'Well yes, but *why* is it that? How has it suddenly changed? Why has this gastric stem cell suddenly changed into an intestinal stem cell? And what are the stimuli for that?'

And what are you discovering? Have you teased out any of the things that are stimulating a stem cell to make something in the wrong place?

No, that's the $64,000 question. The change from one stem cell to another stem cell is the transdifferentiation event we want to understand. We already know *how* it happens, right? We know how the metaplastic cell colonises the gland and becomes an intestinal gland, and how that gland then spreads through the mucosa. We know the names of the genes that are involved, but now we've got to find out what's happening to those genes to make the whole thing change. It's a long process, and I won't complete it by the time I retire. But at least we'll get somewhere.

So in your time as a pathologist, what have been the big technological advances that have allowed you to make these discoveries?

Well, obviously things like immunohistochemistry. When I went to Oxford in the mid-1970s this technique was just being worked out by people there: David Mason and others. Now that's routine. But the thing that has really made these advances possible is molecular biology. DNA technology enables you to dissect out pieces of tissue using what's called a microdissection apparatus, and you can actually extract the DNA and look at the genes you're interested in from these tissues – these paraffin-embedded tissues – and you can find the mutations in them.

When I first began working we never dreamed you would one day be able to get ordinary tissue sections, extract the DNA and look at the molecular make-up. And in specific cells . . . With the techniques we have now, you can dissect out individual cells!

So you can find out where in tissues the tumour started?

In individual cells, yes we can. Yes, absolutely.

Obviously it's very important to understand the mechanisms, but has this led yet to new treatments?

No. I think one of the pleasures of working where I do is that no one is looking over my shoulder and saying, 'Where's the clinical relevance of this?' If my lab was in the medical school, or being funded by different people, they may say, 'Well, how's this going to be translated into treatments?' It won't be, but you can conceptualise quite easily. If you could intervene at critical points through understanding the disease process, you could stop these lesions growing. Now that's a long way off, because we don't know the mechanism of the process that drives this yet. But we will do.

So what motivates you? Is it intellectual curiosity?

It's disease – disease is what's always turned me on, basically, and how the tissues respond and how they try to heal themselves. It's hard to explain. You can say that disease does not exist outside humans, right? So you must therefore be interested in humans? And I say, 'No, I'm interested in the disease process itself.' This would not be a very good thing to say to a modern interviewing committee if you wanted to get into medical school, quite frankly, because they're looking for empathy, and for communication skills, etc. But I think there should be room for people like me in the profession – those who are not interested in people as 'suffering patients', but in the disease processes that they suffer from. Because from that will come eventually an understanding of how these things can be prevented.

Are you as turned on today by the quest to understand disease processes as you always were?

Oh yes. I couldn't live without it. I was talking to some American colleagues, and they were saying that none of their deans were involved in research and how did I find time to do it. I said, 'Well, I couldn't *not* do it!' I can't imagine life without research. No, when they make me close my lab that's the day I'll have to retire, basically!

FORENSICS IN A CULTURE OF VIOLENCE

Patricia Klepp
*Senior Specialist Forensic Pathologist; Senior Lecturer in
Forensic Pathology, University of the Witwatersrand, South
Africa*

The rate of violent crime in Johannesburg and its deep fringe
of poor, overcrowded townships is one of the highest in the
world. This is the territory of Patricia Klepp, the first woman
to qualify as a forensic pathologist in South Africa. In nearly
30 years as a state pathologist, Klepp has investigated many
high-profile killings, such as those perpetrated by Winnie
Mandela's notorious bodyguards in Soweto, and the shooting
of South African Communist Party leader Chris Hani in
1993 that threatened to derail the fragile peace process
between the Nationalist government and the black majority.
'With Chris Hani's assassination . . . it felt like a defining
moment for South Africa,' she says. There has been, besides,
an endless stream of more routine deaths in police detention
and street and domestic violence. The caseload has been
so big that visiting judges from overseas have been known
to query the number of noughts on the record of autopsies
Klepp has performed – often five in one day – thinking there
must be a mistake.

Though she confronts daily the cruelties of life in a
country with conspicuous extremes of wealth and poverty,
Klepp is passionate about her work: 'I would do it even if I
were not paid,' she says. 'I get tremendous satisfaction out

of making findings on the bodies and then going to court and answering questions.' It is interesting to note, therefore, that forensic pathology could not have been further from her mind when she went into medicine. Indeed, she passed out the first time she was confronted with a dead body to dissect at medical school.

———•·•———

I went into that dissecting hall and there was the most horrific smell that goes with those embalmed bodies. So the very first day we started cutting up this body I found myself on the floor, with some of the senior doctors who were training us in anatomy getting me to breathe into a paper bag to revive me. To this day no one can believe that it was Pat who fainted, who couldn't handle the bodies and who now spends all day, every day, with bodies!

So how quickly did you get over it?

If my mother had not already bought me all the textbooks in that second year, I think I would probably have left. What I found very distressing was that the body we had been given was a pauper, and I could not believe that a body would ever not be claimed. This was a man who had cancer of the oesophagus and who had died alone, and I was devastated. But we got accustomed through the year to the body.

The fact that he was a pauper – does the social detail of some of the bodies you see still affect you today?

It does. You know, people say to me, 'How on earth do you cut up those bodies?' And I have to reply, 'They're not actually people I know.'

The youngster I did the post-mortem on this morning, he'd been stabbed in the heart, and I thought, 'If this were in England, it would have been on the news that another youngster had been stabbed.' Knife crime in England is a big issue, but here it's not even mentioned. It's just one more death.

His family was sitting in the car park, on the edge of the pavement, with his pile of clothes, and just seeming to talk to each other naturally, you know? Outside the mortuary waiting to claim the body when I'd finished. As I went in they just greeted me as though it was a normal day . . . And they'd lost this youngster who'd been stabbed – he looked about 22. So yes, I do get concerned. In fact, one of the things I enjoy about my work is going to testify in court, because I feel my job has kept me in touch with ordinary South Africans. I really see the grim side of life. Books and movies and television make forensic pathology look very glamorous, but it's not a glamorous job; it's not. I have a lot of youngsters coming and saying, 'Oh, I want to be a forensic pathologist; can I join you?' And I say, 'Well, first you have to have a passion to become a doctor, and then you slowly but surely do forensics.' I think if you want to be a forensic pathologist it must be a passion that develops once you've done medicine, because it's very much a postgraduate degree – you need to know all the other specialties before you start that.

A lot of people go into pathology because they're fascinated by the mechanisms of disease and so on, but what made you decide on forensics?

Certainly I had no dreams of becoming a forensic pathologist. In my day there were no women forensic pathologists in South Africa. I got over dissecting that body in my second year, and in my third year I looked through the window into theatre and saw them sawing off this man's leg, and I found that quite horrific. By the fifth year we were into the mortuary and doing forensic medicine, and I used to stand right at the back – it was not my scene. But I did well in my fifth year, and then I qualified.

By this stage I was married, and I told my mum, who was financially quite strapped, not to continue paying for me; that I would make a plan. So I went to the university and asked if there were any scholarships or bursaries, and I was awarded a bursary. I didn't take much heed at the time as to who had given it to me – and in fact if it had been 2009 I wouldn't have got it, because it was given to me by the government. They were happy at that stage to help white women. Today it would only be black women who would get help. But the government paid the fees for my fifth and sixth year, by which stage my marriage had not worked. I needed to pay the money back, but I didn't have it.

So I did my internship at Baragwanath Hospital [in Soweto, and famous for being the largest hospital in the southern hemisphere] and I had to find a way to pay back my bursary. It could be in cash or in kind, and the 'kind' was one of four options: to work in a leper colony (which didn't appeal because I pictured my fingers falling off); or a TB hospital (and I didn't fancy everybody coughing in my face); or there was work with the airways, which sounded rather glamorous – I pictured myself flying off to France with somebody ill and getting off at gay Paree – but that job turned out to be sitting at the airport vaccinating people to go abroad. The fourth option was to become a district surgeon – a doctor who worked with the government in the Department of Health but doing a lot for the police.

So I worked as a general practitioner in the city, looking after the police and their families, and doing any sort of state-related work. We would examine people who were going to become policemen; we'd do the disability grants, the old-age homes and the vaccinations. And during that year my boss, Dr Kemp, said to me, 'Pat, why don't you come and see whether you like the work at the mortuary.' So I went across to the mortuary, and I've been there ever since, which is now almost 30 years!

Initially I did what we call here a diploma in forensic medicine, which took two years. I went down to Cape Town to graduate, and there were the two doyens of forensic pathology in this country, Professors Hillel Shapiro and Isidor Gordon, and they told me, 'Oh, a diploma is no good, my girl; you must specialise.' I think it was said tongue in cheek, but I got this overwhelming feeling: I'll show you men! And so I left the district surgeon's and the clinical forensic medicine – which was also the examination of rape victims, murderers, drunken drivers, all that clinical side – and from then I just did the deaths.

I spent time doing two further degrees. I was the first woman forensic pathologist, but since then women have seen that it's a very nice specialty to be in because the hours are very good – I mean, I was still able to watch my sons play cricket. I found the bodies very patient – they would always wait for me in the morning if I was a bit late, if I'd had to do the school lifts!

Was it tough at the beginning? Have you found a glass ceiling?

I obviously had to specialise, which took four years. During that time we did histopathology, chemical pathology, microbiology, haematology and then two years of the forensic side. At the end I went to the College of Medicine for my viva. Normally the vivas would be 20 minutes – you'd be in and out. I opened the door and there were seven men. Now there were only seven forensic pathologists in the whole country and they were obviously deciding whether to take this woman into their domain. It had always been just the men. Anyway, after a couple of hours I came out and I was through!

And you haven't found since that your male colleagues have been obstructive, or the police have been difficult to work with because you're a woman?

Not at all. No, never.

How much has the political climate affected your work? I mean, you grew up through the apartheid era. Your family arrived just after the Nationalist government had come to power, and there have been allegations, both during the apartheid years and since, of corruption in the police, and of the state having its own story about what's happened to people.

Well, under the old regime we were *never* allowed to open our mouths about a thing. You would have had to get permission to interview me from the powers that be. We were never allowed to speak to the newspapers – although I still don't speak to the newspapers because I don't trust them; they never seem to reflect things as they are. But there's far more openness about things now.

We got all the bodies of people who died in detention, people who died during interrogation – the youngsters who were either 'tubed' [the inner lining of a car tyre is tightly applied to the nose and mouth to prevent breathing, as a form of torture] or had plastic bags put over their heads, or who were shocked with electricity. And we still get the odd one now.

How do you manage the emotional impact of that – the fact that the authorities are doing things like that to your fellow South Africans? How do you balance your emotional response with your job as a scientist?

Well, I've never been politically active, so it's never inspired me to spill the beans or take other action. I mean, you would get these bodies, you'd make your objective findings – and objectivity is something I've held on to very dearly – and you'd then be called to court to testify. [My principle has always been to] just tell the court openly exactly what you have found. But the police were *brilliant* at covering up what they had done – we often could not find any marks on the bodies.

Really? Give me an example of something that might have happened at John Vorster Square [the headquarters of the apartheid police].

Well, we would get people who had been interrogated and who would have been given electric shocks. The police knew that if you wet the body before electricity is applied, you don't get a burn. It's like if you're struck by lightning, there's not a mark on the body if you're out in the rain; you just die. But if you're dry – for instance, if you're under an umbrella or a tree – it burns you as it goes in and burns you as it comes out. The water, basically, allows the electricity to run over the body. So what they found with these youngsters was that if they wet them first and then applied the electric clips, there was no mark left on the body.

So what signs would you find inside?

Nothing. So the cause of death would be 'unascertained'. Then you'd go to court and they'd want to know from you what it could be. You were often told that the police had just been talking to this youngster in a cheery manner and he'd had a fit and fallen to the ground, and in order to revive him they'd thrown water over him. In fact, I'm sure they were wetting the prisoners, shocking them, and then they'd fall to the ground. So you'd have a puddle of water with a dead person on the floor . . . They got very expert at it.

I used to get very frustrated and angry in court, because there was an advocate who would defend the police – in fact, he made a fortune out of defending them – and he would come up with this condition that we call a 'prolonged QT interval' and which gives you a sudden arrhythmia of the heart. They put down all these deaths of youngsters to 'a possible prolongation of the QT interval', and one couldn't argue that it could not have been that.

You said that your job has kept you very much in touch with ordinary people and with the conditions of their lives. Have you formed an opinion about why people turn to violence and why life is so cheap?

No, I don't think I have. We get called to scenes of crime and our job is the interpretation of those, and the post-mortem and testifying in court. But I have not been involved with the whole psychological side of why the crimes are committed.

Is it part of your protection that you don't analyse things too deeply? Because otherwise . . .

. . . you'd go mad.

You have brought up your children here – how has your intimate knowledge of the violent nature of this society affected your mothering?

I think I have made them quite cautious. They know. They've even helped me – that's how they used to earn their pocket money. I remember being called by the mines once – they'd had a major accident at TauTona mine: there were 18 dead, and the mine company asked me to do all the post-mortems. So I took my older son, who was a teenager, with me. He didn't actually come into where the bodies were, because bodies have never been his scene, but he certainly helped me put all the equipment out.

My sons see the pictures that I bring home – I've put together PowerPoint presentations with all the gore. And in fact I used to be invited to their school: when they wanted to put the boys off alcohol and things they would call me round. James would say, 'Mum, put that picture in – the one with the pick in the eye!' So they knew all about it. In fact, when I used to replay the Dictaphone (because a lot of the work I do on Dictaphone), and they'd hear the whining of the saw for cutting the heads, everyone would leave the room!

But what gets under your skin? How do you psych yourself up and say, 'This is just a body I'm dealing with'?

Well, it's the fact that you don't see the person suffering. I'm not scared of being dead. I think being dead is actually very easy, though the process of dying is not always that easy. I certainly know how I *don't* want to die: I don't want to burn, for instance.

I think it is much more traumatic for the doctors who work in casualty than for me, because I'm not seeing people with the axe in the head, and in pain. By the time I get my body all that aggro has gone. We have very little contact with the family. I saw that family today sitting in the parking area, but generally speaking we do the post-mortems, and we push off to court, and only then do the public come to identify the body.

Tell me, in nearly thirty years as a state pathologist, have you seen things change much?

Yes, a great deal. You know, our laboratories are struggling today. If I send a sample off for toxicology, for example, and I ask them, 'How long till I get the results?' they say, 'Ten *years!*' So now when I do a post-mortem I wait a month, which I think is reasonable, and then I sign it out and I say, 'Unascertained cause of death; toxicology result not forthcoming.' Otherwise I'd have a pile of dockets from here to the roof.

I used to go to court virtually every day to testify in cases. Now if I go twice in a year, that's a lot. The criminal justice system is not operating as it should. It's very depressing.

How do you feel about your work, even when conditions are difficult?

Oh, very passionate. I would do it even if I were not paid. I get tremendous satisfaction out of making findings on the bodies and then going to court and answering questions, and almost getting the feeling from the accused, '*How* does she know that?' The kind of thing that keeps you going is getting feedback on the judge's assessment of you as a witness. I've had cases go to the Appeal Court, and major findings have been made on evidence that I've given: it's very satisfying.

Have you ever been nervous of appearing in court?

No, not at all. I've heard other people give evidence and I think, 'The judge is not going to know what you're talking about,' because they're using such complicated terminology and phrases. One of the first things I learnt from Dr Kemp was to keep the report straightforward and use terminology that everybody can understand, otherwise you'll spend your entire day in court trying to explain it. So I do enjoy testifying and explaining in simple terminology what it actually means.

Tell me about some of the famous cases you've been involved with. I know you were called when Chris Hani [leader of the South African Communist Party, who was gunned down outside his home in 1993, just a year before the first all-race elections] was killed.

Hani was killed 16 years ago, and we were at the Inanda Club at a birthday party when my phone rang. I was actually asked by the lawyer representing the Hani family whether I would represent them. I said it might be tricky because the government might ask me to do him, and sure enough they did. We did the post-mortem at the Diepkloof Mortuary where I'm working at the moment, and they had to fly a doctor up from the Cape to represent the family.

Hani had been shot a number of times, and any shooting, particularly when it's a multiple shooting, is not straightforward. People always ask, 'Why do you have to find all the bullets?' But it was actually crucial to be able to work out from what distance Hani was shot; whether only Janusz Waluś was involved or whether there were two people shooting . . . The angles, the guns and the bullets: no multiple shooting is ever straightforward.

Hani's assassination was a very big political event, wasn't it? How did you feel when you first got that call at the Inanda Club?

Well, I just knew that this was bad news for South Africa. We didn't need something like that. But it was also a turning point: I think people came to realise that we had a major problem here with political violence and it needed to be addressed. You know, you just carry on living in a society without being too desperately involved in the actual politics until something happens, then people realise: 'We're going to have to change.' With Chris Hani's assassination, when those men came forward – like Tokyo Sexwale [a leading figure in the ANC], who spoke so eloquently on the television – it felt like a defining moment for South Africa.

Were you aware of that when you were first called? Or did you switch straight into scientific mode when you actually had Chris Hani's body in front of you?

Yes, I think so. As far as the politics of this country is concerned, I have attempted, in my personal capacity, to assist anybody with whom I've had any dealings. One of the ladies who works for me here, we've put her daughter through high school. And the other maid has a grandson whom we put through Damelin College: we provided computers for him, and he got a proper job.

I think we've assisted where we've had the ability to, but I haven't been involved in the bigger political picture and gone to rallies and things. I was never a clever enough student to stand on Jan Smuts Avenue with placards. I had to be one of the ones inside working – I would never have got through medical school otherwise. So I suppose you live with something even though you don't approve of it . . . But you know, the other thing is that under the apartheid regime there was no talk ever, of anything. I don't think I ever even *knew* why only the white guys went to the army, and black guys didn't. I mean, you were just in this [bubble] . . . We were aware of the police vans arriving and the police running up your driveway and raiding the servants' quarters, but your parents didn't speak up . . .

And when you went to medical school you didn't think about why there weren't many black or Indian students or . . . ?

What I did know is that when we qualified and we all had to go and stand on the steps of the General Hospital to have the class photograph, it was boycotted by the Indian students. I don't think we had a single black student in our class, but the Indians said, 'If we're not good enough to practise our medicine in that hospital, we're not going to stand on the steps for our photograph.'

Even to this day I will go and vote, but I don't do more. I suppose you just sort of withdraw, because otherwise you'd go mad. We did have a doctor working among us, Jonathan Gluckman, who just did private work. Jonathan and I were very good friends. He would often come and watch me doing a post-mortem, and he would be representing the family or the other side. And Jonathan Gluckman spoke up against the police when we had hangings in the cells or whatever. But those of us working for the state, we weren't allowed to open our mouths. Not that I felt tempted to. I mean, I wasn't concealing anything: we did post-mortems on all the bodies we had. I think there were some that never got to the mortuary – that the police were blowing up and burning or burying . . . But all of those that came to the mortuary, we did. And I was never threatened by the police or anyone, never put under any pressure.

You were also involved in the Stompie Seipei case, weren't you? How did that come about?

Stompie was a little boy who lived in Soweto and who had links with Winnie Mandela. I think he may have been spying for the [apartheid] regime, for the government, and Winnie got wind of it. I don't know if he was actually feeding information to people, but Winnie had a man called Jerry

Richardson who beat up Stompie, and I think stabbed him, and then Stompie's body was tossed into the veld. He was found and taken to the mortuary.

Nobody knew who this body was – I mean, we get lots and lots of bodies that are never claimed. So nobody knew what had happened to Stompie. And then about six weeks later there was some mention in the newspaper about Stompie and Winnie . . . One of my colleagues, Jonathan Cook, who had just joined our department, turned to me at work one day and said, 'I did a post-mortem on a youngster about six weeks ago in Diepkloof . . .' And as he spoke, I said, 'Jonathan, I think we might be talking about the boy that's missing.' So we hopped in the car, went out to Diepkloof, pulled the body out of the fridge, and he looked just like the photograph in the newspaper.

His face was still reasonably okay?

He was markedly decomposed but you could look at the picture and see that it was him. So I asked Jonathan, 'Well, what did you find?' And he said, 'Oh no, the body was so rotten I didn't find anything.' So I said, 'How carefully did you look? Come, let's do it again together.'

Jonathan had just called the death 'unascertained', I think, but looking at this whole thing carefully again you could see that he had been stabbed in the neck. Often when a body is decomposing it's very difficult to tell the difference between bruising and decomposition. But in his particular body, by cutting through into it, you could see the decomposed areas; you could also see the bruised areas.

So I basically compiled that report. I went to the High Court, and I must say that was quite nerve-wracking, because obviously Winnie was there. She had George Bizos, the prominent lawyer, representing her. I was sitting in the passage outside waiting for my turn, and I think it was Paul Verryn, who ran the Methodist Church, who testified before me, and I could hear George Bizos giving him a real run for his money. I don't know if I've got all the facts correctly now, but I think Stompie and some other street children might have been given refuge in Paul's church premises, and Winnie had accused him of sodomising them. I don't think there was any truth in it, but she accused him of sharing his bed with these youngsters . . .

The big story was that this was a 'sting' operation, wasn't it – that Winnie had tried to frame Paul Verryn because she didn't like him?

I think so. Anyway, I almost got a runny tummy sitting there and hearing the cross-examination of Paul Verryn!

And did Bizos give you a hard time too?

I went into court, and they made me run through everything. Then they turned to the defence counsel and said, 'Right, your turn.' So George Bizos stands up, he looks at the judge, he looks at me, he looks down, and he says, 'No questions for this witness.' Not a single question! He just accepted my report.

And how did you feel seeing Winnie Mandela there in court?

Well, the time I really felt aware of her presence was at the Truth and Reconciliation Commission. I was asked to go and testify about Stompie, and she was sitting literally a few feet from me, with *big* sunglasses, just looking . . .

But she's never given me a hard time. Nobody has ever phoned me; I've never had threatening calls. I think what's happened is that they . . . I was going to say 'believe me'. I think they trust me. Because what I've done is to hold very strictly to the principle that 'You will hear what is *there*, not what you want to hear.' So if they employ me, they know that they'll get what's there whether it suits the client or not.

Over your years of doing this work, have you seen the picture, the profile, of crime changing?

Tremendously. In the early days if you wanted to murder somebody you picked up a knobkerrie [a wooden club with a heavy knob on one end] or heavy object, and you coshed them on the head. If there was one shooting in the week, the professor would do it and we'd all go and watch. But one day recently I was at Diepkloof, there were four doctors working and we each had five bodies, and all five of mine had been shot. So I went to the station commander and said, 'You know, five shootings on one day is just too much,' because how we work is we do all five simultaneously. You start by describing everything that's on the outside of this first body, and when you've done that you move on to the second body, the third, the fourth and the fifth. It makes sense to have a variety of cases – a drowning, a hanging, a car crash, a shooting – because if they're all the same it becomes too much to remember which had what. The station commander said to me, 'No, doctor, you mustn't complain, because each of the other doctors has also got four shootings . . .' There were *seventeen shootings* on one day.

How much are you seeing of what one would call gratuitous, calculated cruelty, as opposed to violence like knife and gun fights?

Well, through all the 30 years I've been working, we have had dreadful deaths where people have been trussed up, for instance. In one of my most memorable cases, I was called about two in the morning and there was this white woman all trussed up. The police were taking statements from everybody, and there was Emily the maid under suspicion . . .

When they'd phoned me they'd actually given me the wrong address, and I'd gone to the house next door. But I'd seen people sitting on the patio and I walked up and told them who I was. The dead lady's son was there, and he said, 'When you see my mum, please tell me what you found.' So I went back afterwards, and he said, 'No, the one person it's not is Emily the maid. She's worked for my mother for 20 years and . . .' Anyway, the dead lady was all trussed up, but by taking her temperature and seeing how stiff she was and how settled her blood was I worked out that she'd died shortly after 8 p.m. I did the post-mortem and I could still see rice and peas and chicken in her stomach.

The story went that it was a Jewish family, and she'd had them all there for Friday night dinner, but they'd had an altercation and they'd left early – shortly after 8 p.m. Very security conscious, the lady had let them out at the garden gate, locked that, locked the security gate, locked the door, only to find herself now locked in the house with her killer. Emily told the police that she'd heard intruders – that they'd even knocked on her [Emily's] door and she'd not responded; she'd acted as though she wasn't there. She said she heard them finally drive off at midnight, and she'd heard her employer scream at about that time. When I got to testify in court I said she'd been dead since eight o'clock, so Emily's story of midnight didn't fit in. It turned out to be Emily's son and daughter who had killed her. The son hanged for that death, so it shows how long ago that case was – when we still used to hang people. The daughter got 16 years.

If I hadn't been out there that night and assessed how long she'd been dead, they wouldn't have been caught. Now we are not called to as many crime scenes as we used to be; the police are not nearly as good as they were. That old regime put together a police forensic laboratory in Johannesburg that the world would be proud of. We lost one of our top ballistics guys to America, and he told me that when he submitted his CV and told them how many cases he'd done, they said, 'You've got too many noughts on the end.' He said, 'No, no. It really is that many.'

Do you think all the things you've seen have made you cynical, or hard, or just very realistic about what human beings are capable of?

I think people are just generally quite wicked.

Do you? And did you feel like that even before you started in this line of work?

Not to the same degree. But even with politics, when I look at politics in this country, it is just so tough. But if we all got agitated about it we'd go mad. And you know, I'd rather live in a tough, interesting place than Sweden, or Switzerland, or . . .

When you did your training in forensic pathology, was confronting death and the evilness that human beings are capable of addressed at all?

No, there was no training at all. None at all. I think those who can't cope just wouldn't do it.

As a teacher yourself now, do you ever address this with your students? Do you warn them about the sort of things they will come up against, or do you expect them just to learn to cope for themselves?

Yes, I do address it. One thing I discuss is what to do if somebody close to them were to die. For instance, when my brother died [in a mid-air collision between two small planes], my father went to Pretoria to identify his body. My brother was severely burnt, and my father couldn't identify him, and we had to get a forensic odontologist to look at his teeth. My father was never the same again. Never. So I tell the students: one, the mortuary is not a great place to see; and two, if the person has been severely injured, it's not appropriate for a very close family member to identify the body. It should be somebody one step removed – either an uncle, or someone who can cope, should go and identify them. So one does think of that human side. But no, it's not in the training of the students. It's purely objective, scientific findings that you're taught.

How important do you think it is for families to find out what has happened to a loved one, to have closure?

I think it's very important. But you know, life seems so cheap sometimes. That family sitting in the car park this morning – it made quite an impact on me. They didn't really seem fazed that he was dead. It was almost as though violent death was part of life, you know?

And with Stompie Seipei's body, why hadn't his mother – why hadn't *anybody* – come to the mortuary to look for him? He lay there for weeks . . . And then there were accusations that the government was hiding his body in the mortuary. We weren't hiding him – nobody had even come to look for him. So life seems quite cheap.

But I do think that closure is important for people. I got a call on Friday from the doctor of the family of a Lebanese priest, Lionel Sham, who'd been murdered. The family had asked their doctor to phone me, and though I'm not meant to divulge anything, as I said, I told him everything, because I just think that it *is* very important that people know what happened to him.

Two black youths had killed him. He was in the church, and he always used to let people in who needed help. They coshed him on the head, bound him up and dumped him, then took his car and some of the possessions.

There was another terrible story recently, of a little girl, that you were involved with . . .

Yes, the judge's granddaughter. Her father was a magistrate and her mother, I think, was a lawyer. They'd gone off to work and she'd been left at home with the maid. Intruders broke into the house and tied the maid up. The little girl was missing, and because of her being a judge's granddaughter there was a huge effort to find her – they sent out police helicopters and search dogs and everything. And eventually somebody decided to pick up the bed in the main bedroom, and there was this little girl's body. People said, 'How on earth did the police not find her for so long?' But it didn't look as though a body could be under the bed, so no one had thought of it.

And did you go to the scene of the crime?

Yes. One always feels a bit like an intruder into someone else's home. You put on your white coat and you go marching in through dining rooms, along passages and into bedrooms – it's a real intrusion into other people's space, but you just realise it's a space where terrible, terrible things have just happened . . . It's also fascinating; you feel very privileged.

People say, 'Gosh, you do fascinating work.' And they're right, I do! It is like magic. Some of the things you find, you just can't believe. We've had bodies in car boots at Jan Smuts airport: burnt-out cars where you take a body out that's just charred. You can have a fractured skull from fire, and I had one case where I just didn't know whether the fractures were from the fire or whether the victim had been assaulted. In the stomach was a *tooth*. You look at such things and you think, 'Now how did that tooth get there? He must have swallowed it when his jaw was fractured . . .' These are the kind of things you won't ever learn from a textbook!

One of the things I wanted to ask you is: what are your feelings when you first go into a room or a mortuary with dead bodies? Do you feel there are people there?

They are people, but one doesn't sense that they are people in the true sense of the word. That feeling, call it a soul or a spirit or the *essence*, of a person is not there. It's difficult to explain. It's just like an emptiness. One realises it is just the casing, the shell, of the person. If I go into the mortuary and there are 8, 10 or 12 bodies and I'm alone with them, I really consider that I'm in the mortuary with 8 or 10 bodies, not 8 or 10 'people' who are simply lying there and not talking. As pathologists – I think it's common with all of us – we don't like to work on a body that's warm.

Is the 'essence' still there when a body is warm?

Yes. I felt it particularly with a young Indian girl who had taken an overdose, and as I cut into her I just thought, 'I hope you are really dead.' Because when a body is still warm and rigor mortis hasn't set in, you get this feeling more that the person is still around.

And that little girl under the bed, was she still a little person or . . . ?

333

She had her pyjamas still on. But again I didn't feel that there was a little soul that was agonising and fighting for life. She was actually past the horror, and she was just no longer there.

And does that make it more bearable for you?

Yes, I think so. I know we all are going to be dead someday. I don't believe in ghosts. In fact, I don't even believe in a life hereafter. I just think that's a consolation that's held out to people. I think we're given 10, 20 or 100 years to *live*, and that's it.

Do you think your work has influenced your feelings and beliefs?

Yes, very much. I am what they call a medical referee for cremations, and I go through perhaps 30 sets of papers *a day* on dead bodies. I visit the various crematoria and these bodies are burnt and discarded, and bodies are buried and they're gone, so the whole *body* to me is unimportant. I think the soul is what makes a person when they're alive. When that soul is gone, that person has *died*.

So is the fact that you are doing something worth while – that you're making a contribution to the justice system – important to you?

Yes. I was in court once and I was telling Judge Edwin Cameron, who's always charming to me, about a case I'd had where I was helping the defence. It was a woman who'd had her husband, a headmaster, killed, but she claimed she wasn't part of it.

I mentioned to Judge Cameron that when we get an important murder, I'm sometimes called to the scene of crime to help the police. He said, 'Hold on, Dr Klepp . . . With all due respect, what constitutes an *important* murder?' I said, 'Oh my God, my Lord, they're all important!'

That drew you up sharp, did it?

Absolutely! But quite honestly, it *is* the 'important' murders. When it's the judge's granddaughter, there are helicopters in the sky and sniffer dogs. But when it's not the judge's granddaughter . . . So, he was right asking the question, but in fact I'm correct in saying the 'important' murders. I had to say, 'Oh no, they're all important', but we all know that's not the truth.

LESSONS IN LIVING AND DYING

Francisco González-Crussí
*Retired Professor of Pathology, Northwestern University
Medical School; and Head of Laboratories, Children's
Memorial Hospital, Chicago*

Francisco González-Crussí has done as much as anyone to bring the practices and preoccupations of pathologists – so often known as 'the backroom boys' because of their hidden role in the care of patients – into the public arena with his wonderful collection of medical essays, which started with *Notes of an Anatomist* in 1985. Born into a very poor home in Mexico City, he had an enquiring mind and unwavering support from his widowed mother. He qualified in medicine in his home country before going to North America for specialist training in paediatric pathology. For nearly 30 years González-Crussí was head of pathology at the Children's Memorial Hospital in Chicago, where he pursued his passion for literature and for writing alongside his career as a remarkable diagnostician, doctor and teacher.

González-Crussí, more than most, explores what his work as a pathologist has taught him about the meaning of life and death. 'The contemplation of death – the spectacle of the cadaver being opened at dissection – is truly an important experience,' he says. 'You can't help but say, "We are all made of that stuff . . . That's going to happen to me too." [This is] the understanding that touches the heart.'

I was born and raised in Mexico City in a family of rather small means. My mother was a widow and had two children to put through school, and it was very difficult. It was a ghetto-like neighbourhood. But somehow I was fortunate and was able to pursue my studies. Probably I'm atypical in that I grew up with a double line of preference: on one side was literature (that's why I write), and on the other was medicine.

I came to a crossroads where I had to decide whether to go into the school of letters and philosophy, or the school of medicine. But given my background, the suffering of my mother that I could see daily, I thought that going for letters, especially in a third-world environment, was probably too idealistic. On the other hand, medicine, biology, is a fascinating field, so the choice was clear.

What sort of schooling did you have, and how easy was it to fulfil your ambition of studying medicine?

It was not easy. But what was particularly stimulating was the presence of role models. I had a very inspiring biology teacher in high school, and later on in medical school. At that time Mexico benefited tremendously from the diaspora of Spain. Spain was under the Franco yoke; the Civil War had just taken place, and the cream of the Spanish intelligentsia had emigrated. They naturally gravitated towards the Spanish-speaking areas of the world and many came to Mexico. Spain's loss was a great benefit to Mexico.

With the emigrés came a very inspiring professor of pathology who had trained in Germany, because at the time – the first half of the twentieth century – the Mecca for pathology was Germany. After the Second World War, the United States became the leader in this speciality. But this professor, Isaac Costero, was witty, his lectures were heavily attended – even those who were not officially his students crowded into the room. At the same time he was erudite and wise. I said, 'I would like to be like that man!' He was truly a role model.

So was he the person who inspired you to do pathology rather than clinical medicine?

He was a major force in that respect. Then later there was another. One of his students was a young man called Ruy Pérez-Tamayo. Dr Costero had all these baroque classifications of tumours, with all kinds of complicated terminology taken from the German pathologists that he knew so well. His student had actually trained in the United States, and he thought much of that was burdensome and useless knowledge; that it was not so important to know all these fancy terms; one had to have a more dynamic concept of pathology. Pérez-Tamayo had an experimental turn of mind, and this was a revelation to me. Suddenly I saw that there were two worlds: the old world of the erudite, complex classifications, all of which sounded very elegant, very impressive; and another which saw through this tangle of things into what was more essential, more rational.

There was friction between the student and the teacher, but it's the fate of all great teachers to see their pupils surpassing them! Ruy Pérez-Tamayo, now in his late eighties, is one of Mexico's leading intellectuals. These were the two major forces in my life, the two people that I wanted to be like: erudite, knowledgeable, like the first one; brilliant, enquiring, and even handsome like the second . . . He had *everything*! [*laughs*]

Was your mother supportive of your studies?

I was very fortunate – I knew from early on what maternal love really means. My mother sacrificed herself for me, and I appreciated that. On the other hand she did not quite understand what I was doing. To her, after all this long sacrifice, I was supposed to see patients; become the family doctor of the neighbourhood or something. And I didn't want to see patients; I wanted to be in the laboratory and to study the theoretical aspects of pathology. She didn't quite understand that.

There was another incident that I think was very important, although it is not directly related to medicine. The French have high prestige in Latin America, and they had a very wise policy of giving scholarships to young people to spend some time in France. I was only 16 or 17 when I participated in a competition that was actually a proficiency in French language, and I won. My mother could not even have dreamed of sending me to Europe to study, but through the good graces of the French government I was sent for a few months to Paris.

My mother scrounged what she could and gave me what would be about $40 today. The ticket was paid by the school. I had to go from Mexico City to New York, then take the boat to Le Havre, and the train to Paris. Never having been away from Mexico City, and all of a sudden having to do this, it was an eye-opener! I developed then a love for French literature – as I think you can see. [*he gestures towards a pile of French books*]

Before your scholarship to France, were you aware of a world 'out there' that you would like to join?

Through literature, yes. Another very fortunate thing for me was that, in the very poor neighbourhood where I grew up, there was a Jewish family next door who had fled the devastation and persecution of the Second World War in Europe. Jewish families have this emphasis on academics, and these neighbours were readers. My mother was working all the time, so I gravitated towards this family. I was almost like an adopted child to them. I could see their interest in reading and that helped me very much, I think, because the other kids in the neighbourhood were not like that at all: they were overwhelmed by all the negative influences of poverty and destitution. But this family had the experience of having lived in some *shtetl* in Poland, and having to survive somehow, and they brought with them this tradition of, 'You respect the man who *knows*.'

So as a 12-year-old, talking about these things, reading along with them; then in high school, the teacher of biology; and then later in medical school, the professors – all were inspiring . . .

Another aspect in how my life worked out is my own personality. In everyone there are the outside influences and those that are inherent in the person, and I probably did not have the emotional fortitude to get into the world of medicine. I was extremely shy. When I started the rotating internship during medical training, I was *terrified* by the responsibility of caring for the patients. Now in a properly structured teaching programme the professor should always be there to back the beginner, and one should not be left alone with the responsibility. But this was North America – I was an intern in some clinic in Colorado – and sometimes the interns were left alone. To me it was an unbearable emotional burden: 'What if something I do has bad consequences?' I started losing my hair because of the pressure under which I was living.

At the same time I loved medicine. It was *extremely* interesting – the pathophysiology, the origin and evolution of a disease. I wanted to study medicine, but at the same time

I was a little afraid of *practising* medicine. So, remembering the example of the professors that I most admired, I thought, naturally, 'Pathology is my field!'

I had started doing pathology in Mexico. I worked as an assistant to the professors who inspired me in various hospitals in Mexico City. It was the pathology of poverty, and I remember seeing so many patients, or autopsies, with amoebic abscess of the liver – something that one does not see in North America, except in immigrants – and tuberculosis, which was very common in Mexico at that time.

When you first started pathology in Mexico, how was it taught?

By the time I was there the mentors had mostly been trained in North America, so the systems were very good. The resources of the institutions were more limited because of the poverty of the Third World, but the systems were good: the teachers knew what they were doing, and they had the correct literature. I remember reading avidly the American journals of cancer and so on, because they encouraged us to do so.

The shift occurred in the 1950s and '60s; many of the people who had trained in the US came back to teach in Mexico. Before that it was mainly the European influence. As I say, my Spanish professor had trained in Germany. He used to have in his office *The Henke–Lubarsch Handbook of Pathology*. And he used to say that if it was not in Henke and Lubarsch, it didn't exist! But then the young people came with completely different ideas of how pathology should be studied.

Tell me, before you went into pathology what were your experiences of death?

Well, I remember my father's death when I was 10 years old. But he didn't die at home – he wanted to be taken to his home town, which was miles away from Mexico City. Deaths in the family were moving experiences, but I didn't really think much about them. At least, they didn't influence me in my choice of career.

I remember an experience with a journalist who had already decided before coming to interview me that I must be somebody with a morbid turn of mind – partly because of the books I had written, and partly, as she said, because 'otherwise, why did you choose to be a pathologist?' She had this idea that the pathologist had to be some revised version of Dr Jack Kevorkian [the Michigan pathologist who went to prison for assisting his desperate patients to commit suicide], someone constantly obsessed by death. But it was not like that at all. I went into pathology because it is intrinsically interesting, and because it is really the foundation – one of the sustaining pillars – of medicine. And secondly because of the personal role models that I had. But an obsession with death? No, I don't have it.

And yet, by the very nature of your job, you are dealing with the fact of death, aren't you?

Well, you know, the contemplation of death – the spectacle of the cadaver being opened at dissection – is truly an important experience. I once had someone ask me, 'You have spent your life doing autopsies (not nearly as many as they imagined, but I did do a good number!) – what pearl of knowledge have all these experiences with the dead taught you?' I didn't know what to say, because it was one of those loaded questions like, 'Tell us what you know of the meaning of life . . . in the next 10 seconds'!

I said, 'Well, I think it has taught me the tenuousness of life.' Because I recall people dying from such banal causes – a healthy 21-year-old asphyxiated by an olive; a four-year-old, angelic little girl, who collapsed in the physician's office after an injection of penicillin, and many others like that. So I told him, 'Perhaps if I have learned one thing it is that our life is suspended by a thread.' And he said, 'Well, in that case you really didn't learn anything, because that we knew already.' That's what he said!

Later, reflecting about it, I thought, 'He didn't really understand what I was saying.' That life is finite and that we are all going to die: yes, that we know. But we know it from a purely intellectual standpoint. But when you actually look at it – look at the evisceration of a cadaver – it's like somebody dragged you to contemplate the spectacle of your own dissolution. You can't help but say, 'We are all made of that stuff. All that entanglement of cogs and wheels I see, that's the same as I have here [*he touches his own body*], and now it's just a lump, an inactive mass of proteins in decomposition already. That's going to happen to me too.' To actually see it, then you understand it in an affective way. It's in the terrain of the emotions that you are now, the understanding that touches the heart. The other was purely intellectual. I think my familiarity with dead bodies gave me a closeness – a great sense of proximity to the fact of death, which otherwise one tries to keep at a distance.

On the other hand, we human beings are probably not made to have this sense of the proximity to death – just as we are not made to perceive the daily functioning of life. We never *feel* the stomach unless we have indigestion. We never think of our brain unless we have a headache. If it is not diseased, the heart goes on with its rhythm, and we never realise it is there. And just as we are not made to perceive life, we are not made to perceive death. When it happens, it happens, and I, like every human being, tend to forget about it.

Yes, but it is the sort of thing you do ponder in your essays, isn't it?

At the beginning I was a little mischievous. Because the books were successful, I thought, 'There is a great interest in death. And since nobody knows, I can pose as an expert!' But then I realised I was creating the impression that pathologists must be death-oriented, and I didn't like that. I wanted to show that there are so many other things, so I've written a book about birth, a book on love . . . But when I'm invited to speak, that is the question that always comes up: 'Being a pathologist, what about death?'

So these other aspects of pathology that are to do with understanding disease to help the living – what have you been involved in?

Well, my career is now over because I have retired. But in general the pathologist has several choices. He can be diagnostic-oriented, which in itself takes a whole lifetime because the explosion of knowledge has been so great. If one becomes proficient at diagnosing, at reading biopsies, it is very satisfying, and you know you have done something for the patient. It used to be that all you could do was look at the biopsy. You had a repertoire of images in your brain, and you identified that this tumour was A, B, C or whatever, and that was it. Now there are many other things that can be done besides the appearance of the lesion under the microscope. There are tests of molecular biology; all kinds of things that contribute to a diagnostic opinion, and they should all fit coherently, all be part of the puzzle. So that's one avenue for the pathologist

– to become an expert diagnostician. That's mainly what I did.

Then another thing is teaching, which I did, naturally. You sit with the students and look down the microscope. The thing is, when you look at something, you don't see it unless you *know* what you are seeing. The eyes are really guided by the brain: you must know what to look for, otherwise you don't see it, and there are many examples of that.

An English author discussed in a book how it would be if you could put yourself in the shoes of the first anatomist – when you lift up the breast plate and look at the organs, you know? They look like a mass, a continuous mass. You know *now* that if you cut this fibrous tissue there is going to be a heart, but the first people would not know even where to cut. And it was very difficult to discern one organ from another. The point is that you only see that which you already know, and which you are *prepared* to see. And in teaching pathology, you sit with a student and you teach him to see. Then after that he's on his own, and he depends on his own ability.

Some people 'have the eye', but there are others who may have tremendous abstract reasoning, and be excellent in many aspects, but they don't have a 'visual intelligence', so they don't become adept at microscopic interpretation. But in addition to visual skill in interpreting, the first thing you need is to like what you see. Some of those microscopic fields have an intrinsic beauty to them, and some of it is just *gorgeous*. When you have fluorescent microscopy, the cells with the nuclei like shining stars against the dark background, it's a beautiful thing.

So diagnosis was your focus, was it? And where did your career take you?

A MATTER OF LIFE AND DEATH

I specialised in paediatric pathology, because I came to the United States as an immigrant and that's where opportunities opened up. I think I would have been happy in any branch of pathology, but it so happened that paediatric pathology, when I first came, was really a developing field. There were not more than 100 people in the whole country practising that speciality. They had what was called a 'paediatric pathology *club*', because it was just a group of friends, really. Now it's a 'society' with hundreds of members. Hundreds is still not thousands, but it is much better.

Why was it so slow to develop?

Because people thought that children were just adults in miniature. People, when they studied, knew that it was not the case, but there was this implicit assumption that, on the one hand, yes, there are such things as congenital anomalies – stillbirths, babies with hearts all twisted, absence of intestines or whatever – but the attitude was, 'You can't do anything about them, so you're not helping anyone by diagnosing those abstruse syndromes.' There was a general lack of interest, I think, for that reason.

Then in the 1940s and '50s there was a pioneer in the United States: Edith Potter. She was really the founding mother of paediatric pathology. She was working in Cook County Hospital, Chicago, and she was skilful enough to obtain permission from the city authorities to autopsy any stillborn child there, and she did fantastic work. She really made what became later like the bible of paediatric pathology: an inventory and a categorisation of the different anomalies that can be present.

So she put paediatric pathology on the map, did she?

Yes, she did. I got involved not just in the area of neonatology, but in the wider field of paediatrics, which extends up to teenagers as well. I was doing mostly surgical pathology, so I was looking at tumours – and again, the tumours that children get are different from those that adults get. I wrote a couple of now obsolete medical books. One was on tumours of the kidney in children, which are dominated by what is called a Wilms' tumour, or nephroblastoma. The other was on teratomas, which are fascinating. These are tumours composed of all kinds of tissues exotic to the place where they arise. For example, in the ovary you find thyroid tissue, hairs . . . *hairs!* There shouldn't be hairs in the ovary, so the first people who ever saw a teratoma – with bone and thyroid and hairs – in pre-scientific times, thought that that woman had had carnal contact with the devil or something like that.

Where do teratomas come from?

It's a not uncommon tumour of children, and some of them are congenital. They tend to grow in the middle of the body. In other words you see them in the neck, in the mediastinum [the area between the lungs], in the genital organs. In fact, they are most common in the testicles of the male and the ovaries of the female. If a baby is born with a teratoma of the testicle, it usually is of no concern: it is removed by the surgeon and nothing happens. But in an adult who happens to have a teratoma of the testicles, it is a bad business. Usually it has a malignant component, and a rapid course of invasions and metastases. So age seems to be a very strong determinant of the prognosis of the teratomas of the testicles. They have, as I say, different locations. Congenital teratomas are most common in the coccyx. Babies are born with a huge bag of exotic tissues in the lower back. It's a horrible thing – some of the pictures are horrifying.

Where does a teratoma start?

Well, the classical theory is that it must come from germ cells. Early in the embryo you have cells that differentiate to become bone, brain, gut and so on, but there are some cells that remain primitive. Why? Because those are the ones that give rise to the gonadal cells. These are the germ cells (from Latin *germen*, meaning 'seed'), and they are called 'totipotential', because they have the potential to become anything. Those very primitive cells presumably become trapped in the organism and are the origin of the teratoma: if they become tumoral they have the potential to develop into all kinds of tissue.

So were you pleased to have found yourself directed towards paediatric pathology when it was still a developing field?

Yes, yes. When I really got into it, I realised it had been neglected for no good reason. People thought, as I say, that some diseases are so far advanced by the time the child is born that there was no point in studying them. But with the upsurge in genetics and so on – and not only genetics, there are now people who are even doing fetal surgery: they operate on the baby and put it back in the mother's womb! – there is now interest in all kinds of paediatric pathology.

I was in the Chicago Children's Memorial Hospital, and we used to get referrals from all over. It used to be that at certain ages, the second most common cause of death, after accidents, was cancer, and I saw a lot of those specific childhood cancers that I've mentioned.

And did you work directly with the surgeons?

Yes. We had a personal rapport. We knew the surgeon because we had seen all his cases; we had helped him. We were called into the operating room, and sometimes he'd even say, 'Look in here. What d'you think this is?' while the patient was being operated on. 'Where should I take the biopsy from?' It was beautiful: an experienced pathologist working with an experienced surgeon and having a harmonious relationship is in the best interests of the patient.

In my time, the place where you processed the piece of tissue that the surgeon gave you was next to the operating room, and you had to put on your scrub suit and mask like you were a surgeon, because your room was within the operating area. The nurse or you received the piece of tissue from the surgeon's hands, and you went into the next room to do the processing. Sometimes the surgeon would come out to see what you were looking at down the microscope.

Now, they say there was too much wasted time; pathologists have other things to do. And now I understand that we're not called into the operating room. There is a television camera, and the pathologist is in another part of the hospital. You look at things on the monitor and talk to the surgeon by microphone. You make your sections and you report it; it's impersonal. Perhaps like every old person, I am out of sync with the technological advances; I belong to another era. But somehow I think that when you lose the personal contact, you lose something else.

So tell me, what percentage of your working time was devoted to diagnosis in living patients and what to autopsy?

Unfortunately, at the end of my career, administration took a large part of my time. Before that about 60% of my time would be in service, 30% in teaching and 10% in administration. Autopsies were part of the service. And that, as you probably know, has been decreasing gradually – not so much in paediatric pathology, because in paediatrics there are so many things still unknown that even the physicians feel the need to request autopsies. But in adults, I understand there are hospitals where less than 1% of patients who die go through autopsy.

There are many reasons for that. The main reason, I think, in North America, is the money, because an autopsy is expensive, and who's going to pay? Not the survivors of the patient, because they see no benefit – and especially when they are still grieving, it's hard to ask them to pay. So the only way is for the hospital to absorb the cost, and today, especially in North America, the health system has become a health industry – they are competing for patients. The situation is, I think, rather shameful; it's become like a business.

What do you as a pathologist feel the decline in autopsies has done to your profession? How badly affected is it by an industry which says, 'We can manage without that'?

The effect is negative, I'm convinced. As I say, one of the main causes of the decline in the autopsy is economics, but there have been other causes. One is the fear of being sued by the patient's family should the autopsy find something that the physicians didn't see. The arrogance of the physicians is another cause. They say, 'In this day and age, we have such technological advances that we already know what the patient had.' This has been proven incorrect by serious studies, some published in the *New England Journal of Medicine*, where in spite of all the advances in technology, there were significant numbers of cases where things that they didn't know were uncovered by the autopsy.

Not only that, there are new diseases that can only be known by securing tissues and doing all the studies that an autopsy allows. There are new diseases created by the physicians with the new therapies! New prostheses and organ transplants are giving rise to pathology that didn't exist before, and how can it be known? Not only by biopsies; you really have to examine the whole organs.

Coming to the US as a young immigrant from the background you did, how easy was it to move into a completely different social milieu?

It was a Calvary! First of all I came, so many years back, to a place the interns, *especially* the interns, were badly exploited – underpaid, overworked and not really taught well. But you knew it was leading to something you wanted, so you had to subject yourself to all kinds of vexations.

I remember I got something like $50 a month, plus meals and a room to stay. And they took money for the room. I never had any clothes; I just used the jump suits of the operating room for the whole year. They assigned me, when I first came, chores that a nurse's aide would be doing – washing patients and all that.

The hospitals that received the immigrants, the foreign medical graduates, usually were not the best. The institutions where the standards were highest usually had mostly North Americans. In fact, that was how you got an idea of the quality of the training programme – if you saw that most of the resident staff were foreigners, then you knew that was not such a great place to be. But I had to go there because I did not come with a recommendation from a professor in Mexico, as I should have done.

So I had years of wandering around with a sense of frustration, and feeling, 'I should not have come here. I could have learned just as much in my own country. I'm wasting my time!' But eventually I did get into a university hospital, which is what I wanted, to break into academic medicine. And then I did try to go back to Mexico. But by then there were not the right opportunities. The best places were taken already by people who had done it the right way – who'd been recommended by professors to the United States and then received back to good jobs. They had no place for me.

I didn't know what to do. And then someone said there was a job in Canada, would I go to Canada? Well, I would go *anywhere*. 'But it's in paediatric pathology.' 'That's fine.' So I went to Canada, to a university hospital; an excellent place, very conducive to study: Queen's University, in Kingston, Ontario. They had a lot of resources, and excellent pathologists. And then I came back to the United States and for various reasons that are not germane to this I never went back to work in Mexico.

Tell me, during your childhood, what did your mother do to support you?

Well, my father had a small drugstore. It was like the ancient apothecary shops, a tiny little corner shop where we sold all kinds of traditional medicine. When my father died my mother had to take over. She had to avail herself of the services of a pharmacological chemist, who came every month to get the salary and didn't do anything! But that was our daily living; we got money to eat from my mother working in that little drugstore.

Some of the things we sold, you wouldn't believe. For example, in Latin America, particularly in Mexico, there is a pathological condition which is called *susto*. *Susto* means 'frightened, startled', and to be startled is to be *asustado*. In our neighbourhood there were many things to be startled about: a brawl in the street; the police chasing somebody;

a fight between husband and wife and screams and all that; somebody hit by a car. People would come to the drugstore agitated and trembling and say, 'We need something for *susto*.' So we would give them something. It was just a placebo, but it worked!

My mother sometimes wanted me to be in the drugstore to help her. Poor woman, she worked from morning till night – the drugstore closed at 11 p.m. – and the least I could do after finishing my homework was to help distribute the medicine, or whatever. For *susto* it was magnesium carbonate, a little cherry syrup to make it tasty, and distilled water; you stirred it and gave it to the people. They took it right there.

But did you, too, believe in susto, *growing up in that environment?*

No. It was just a job, and as I was making the cherry mixtures I didn't think about it. I just thought about getting back to play or to do my homework!

My mother must have been in her sixties when she quit. By then I could send money from here to help her a little, and she could retire. In the end I bought my mother and my sister a little house. So, that's my personal situation. I reflect upon it and, well, some things did not come out the way I expected, because ideally I would have wanted to go back to Mexico. It's a difficult thing to leave your home, your country, your language, your traditions, your friends, for a completely new place. It's always traumatic. But as time goes by it becomes more difficult to go back home. Life becomes more complex.

Finally, we talked about autopsies earlier on, but what is your personal experience? Do you see it primarily as a scientific enquiry, or is the body still a human person?

When I am doing an autopsy for the purpose of writing a report and talking to my colleagues, I have to confine myself to the purely medical aspects. But I could never, in the back of my mind, not think of the circumstances that surround that particular individual. That's probably one of the reasons why I wrote my essays: to do the autopsy in the morning, and then to come home and reflect on that case in the evening. I don't know how common this is among pathologists, but a colleague of mine, Dr Robert Bolande, who's now passed away, said, 'Well, you do that because of your background and your sensitivity. You like literature and you read. Most pathologists', as he put it colourfully, 'might as well be working in the Kraft Company.' You know, cutting the meat! They don't think beyond that.

In one of my essays I talked about how the pathologist is supposed to find the cause of death. But what we actually look at is just a very small part of that enormous puzzle, which is the immediate, pathophysiological mechanism of death, period. Behind that are all kinds of social and cultural circumstances. The patient died of cirrhosis of the liver. Okay, he got cirrhosis because he was an alcoholic, and he was an alcoholic because he was reared in circumstances that led him to drink. It becomes a labyrinth; you get lost in the meanders here and there. So I said in my essay that if you were really to write a full report about the cause of death, it would be like Cervantes' *Don Quixote*, where he starts with one narrative and encounters a personage who tells another narrative, and you end up with a novel within the novel. So my autopsy report, if it were to be more faithful to reality, would be a big tome that no one would want to read!

And does it really tease you, the fact that you are not looking at the whole picture?

To me, yes. I know that you have to establish limits: you are a pathologist and you are expected to render a service, a report. I know the other aspect is very important, but that is not part of the report, so if I have the restlessness of putting it somewhere, I put it in a book of essays.

In one of your essays you describe various rituals you had when doing autopsies to show respect to the body. You talked about covering the face – was that something you developed?

No, that is done in many places, and I think it is done spontaneously. And I think it's interesting, because when someone is dead, in reality it is no longer a person. As I say, sometimes it is a mass of protein that has already started decomposing. That's all it is: there are no projects, no emotions, it is insentient. But the appearance is of the person who was alive, and that is mainly because of the face. So you prefer to cover the head before doing anything to the 'person' you know is really not a person any longer, because if you didn't it would be restituting the condition of humaneness that had been lost through death.

ACKNOWLEDGEMENTS

For the chance to wander into and discover a fascinating world that would normally be off my beaten track, I wish to express my gratitude to the Pathological Society of Great Britain and Ireland, who commissioned this book and gave me good advice when asked, but complete professional independence in all matters. Who could ask for a better deal? Thanks are due in particular to Peter Hall, my first guide to this new world, who introduced me to the PathSoc; to David Levison for his enthusiasm and quiet faith in this project and in me; and to Roselyn Pitts and Julie Johnstone for their always gracious assistance with travel arrangements and admin – the kind of background support that allows an assignment to be lots of fun rather than fraught with stress.

I have been lucky to work with Carol Pope of Dundee University Press, a lady with great creative imagination who loves nothing better than to see ideas blossom; and with Anya Serota of Canongate, who has shown equal enthusiasm for this project. I am specially lucky, too, in my agent, Natasha Fairweather of A.P. Watt, who is brilliant at pulling together the disparate strands of an idea into a practical proposition that will keep bread on my table.

Sincere thanks are due to my family – my partner Fred, my mother Mary, sisters Jane and Julie, sons Matt and Lawrence and daughters-in-law Caroline and Valerie – for their constant moral support and willingness to listen to my writer's tales, and for occasional invaluable, professional opinions from the medics among them.

My thanks, too, to Elizabeth Garrett, who provided the perfect environment – Cliff Cottage on the ragged Aberdeenshire coast and lots of TLC – for putting the finishing touches to the book. And to Tim and Olivia Marchant for the same in Wiltshire at earlier stages of the project.

Finally, I owe an enormous debt of gratitude to the women and men who have been prepared to share their life stories with me, and to trust me with information that is sometimes painfully personal and often scientifically tough. And to my friend Olivia Bennett, who, as an oral historian, has read every transcript, given terrific help and advice with the editing and selection process every step along the way, and come to know and care about these stories almost as much as I.

This book is dedicated to the memory of my late father, Peter Abbott, whose career in tropical medicine took us, his family, to homes around the world, and gave me a fascination for all things medical, as well as incurably itchy feet.

GLOSSARY

Adenoma: A benign or pre-cancerous tumour typically of glandular origin.

Anatomical pathology: Study of the structure and composition of abnormal, diseased or injured tissue (*see also* **pathology; surgical pathology**).

Angiography: A special form of X-ray examination that shows the blood flow in arteries and veins.

Antibody: A protein produced by the body's immune system that recognises and attacks foreign substances.

Antigen: Any foreign substance or organism that stimulates the body's immune system to produce antibodies and cells that react specifically with it.

Antiretrovirals: Drugs used for the treatment of infection by retroviruses, primarily HIV.

Autoimmune disease: A condition in which the body mistakes its own tissues as foreign and directs an immune response against them.

Benign: In medical usage, benign is the opposite of malignant. It usually describes an abnormal growth that is stable, treatable and generally not life-threatening.

Biopsy: The removal and examination of a small sample of tissue from a living body for diagnostic purposes.

Bronchioles: The smallest branches of the airways in the lungs. They connect to the alveoli (air sacs).

Carcinoma: A type of cancer that starts in epithelia, the tissues that line or cover most body organs. At least 80% of cancers are carcinomas (*see also* **leukaemia, lymphoma, sarcoma**).

Carcinoma *in situ*: Literally, 'cancer in place', which means the tumour is non-invasive, and has not spread.

Chromosome translocation: A type of chromosomal abnormality in which a piece of a chromosome has broken off and attached to another chromosome.

Clinico-pathological: Pertaining to both the signs and symptoms of a disease (the clinical picture), and the nature, causes and consequences of the disease (the pathology).

Congenital: Present at birth but not necessarily hereditary; acquired during fetal development.

Connective tissue: Serves a 'connecting' function, supporting and binding other tissues. 'Dense' connective tissue includes tendons and ligaments. 'Loose' connective tissue, distributed throughout the body, serves as a packing and binding material for most of our organs. Bone and cartilage are considered to be 'specialised' connective tissues (*see also* **endothelium, epithelium**).

Cytology: The branch of biology that studies the structure and function of cells.

Cytoplasm: The contents of a cell, outside the nucleus.

Dermatopathology: A subspecialty of surgical pathology concerned with skin diseases.

Differentiation: The process whereby an unspecialised early embryonic cell acquires the features of a specialised cell, such as a heart, liver or muscle cell.

DNA: Stands for deoxyribonucleic acid. This is the material inside the nucleus of the cells of living organisms that carries genetic information (*see also* **RNA**).

Electron microscope: A type of microscope that uses electrons to illuminate a specimen and create an enlarged image. It has much greater resolving power than a light microscope and can obtain much higher magnifications.

Electrophoresis: A laboratory technique that is used to separate molecules such as nucleic acids (DNA and RNA) or proteins on the basis of size, electric charge and other physical properties.

Embolus: Something that travels through the bloodstream, lodges in a blood vessel and blocks it.

Endoscope: A flexible tube with a lighted camera attached that is used to view the digestive tract and other internal organs non-surgically (hence endoscopy; endoscopic).

Endothelium: The thin layer of cells that line the internal walls of blood vessels and lymphatic vessels.

Epithelium: The layer of cells covering most of the body's structures and organs, internal and external. It includes the skin.

Forensic: Relating to the application of scientific knowledge to legal problems and legal proceedings (hence forensic anthropology; forensic pathology).

Gastroenterology: The branch of medicine that deals with the digestive system and its disorders.

Genetic sequencing: The process by which the exact arrangement of the units of information on a specific stretch of DNA, or a gene, is determined.

Genome: The complete package of genetic material for a living thing, organised in chromosomes. A copy of the genome is found in every cell.

Haematology: The branch of medical science concerned with the blood and blood-forming tissues; haematopathology is concerned with diseases of the blood and blood-forming tissues.

H&E: A standard tissue stain that all pathologists and labs use. The letters stand for the chemicals haematoxylin and eosin.

Hernia: A general term referring to a protrusion of a tissue through the wall of the cavity in which it is normally contained.

Histology: The study of cells and tissues, usually carried out with the aid of a microscope.

Histopathology: The study of diseased tissue and cell samples under a microscope.

Immunohistochemistry: A technique that uses antibodies labelled with fluorescent or pigmented dyes to identify, or indicate the presence of, specific proteins in tissues when looked at under the microscope.

In situ hybridisation: A technique allowing scientists to identify particular DNA or RNA sequences while these sequences remain in their original location in the cell.

Leukaemia: Cancer of the white blood cells, which are a vital component of the immune system (*see also* **lymphoma**).

Lumpectomy: The surgical removal of a small tumour, which may or may not be cancerous.

Lymphoma: Cancer originating in lymphoid tissue, a key component of the body's immune system. Cancers of lymphocytes (lymphomas) and other white cells in the blood (leukaemia) together account for about 6.5% of all cancers.

Malignant: In medical usage malignant means cancerous; and it can spread to other parts of the body.

Medical examiner: The term for coroner in the US. To qualify as a medical examiner, a person must have an MD and be licensed as a pathologist. A coroner needs to be qualified in either law or medicine.

Mesenchymal cells: Embryonic cells that give rise to various connective tissues such as cartilage, bone, muscle, and lymphatic and blood vessels.

Metaplasia: The transformation of adult cells from one tissue type into another tissue type.

Metastasis: The spread of cancer cells from the original site to other parts of the body (hence metastases: secondary cancers).

Molecular biology: The study of biology at the level of the molecule, the smallest structural unit. Molecular biology chiefly concerns itself with understanding the interactions between the various systems of a cell, including the interactions between DNA, RNA and protein synthesis (hence molecular; molecular genetics).

Monoclonal antibody: An antibody produced in the laboratory from a single clone of cells, which is therefore a single, pure, homogeneous type of antibody that recognises only a single, specific antigen, or protein (*see also* **antigen**).

Morphology: (1) The form and structure of an organism or part of an organism. (2) The study of the form and structure of organisms.

Mucosa: Another term for mucous membrane.

Neonatology: The branch of paediatrics that deals with the diseases and care of newborn babies.

Neuropathology: The pathology of the brain and nervous system.

Oncology: The study of cancer.

Pathogenesis: The origin of a disease, and the chain of events leading to that disease.

Pathology: The scientific study of the nature of disease and its causes, processes, development and consequences.

Pathophysiology: The functional changes associated with disease or injury.

Perinate: The term applied to a baby in the period shortly before and shortly after birth, variously defined as beginning between the 20th and 28th week of gestation and ending the 7th to 28th day after birth (hence perinatal).

Peristalsis: The rippling motion of muscles in the digestive tract (hence peristaltic).

Plasticity: The phenomenon whereby, under certain conditions, tissue-specific adult stem cells can generate a whole spectrum of cell types of other tissues.

Pluripotent: Pluripotent means 'with many powers'. In cell biology it refers to a cell that is able to differentiate in many directions and develop into any of the major tissue types.

Resident: The US equivalent of a registrar in the British system.

RNA: Stands for ribonucleic acid. RNA, like DNA, is found in every cell of every living thing on earth. The relationship between the two, in summary, is that DNA makes RNA, and RNA makes proteins. In other words, DNA is the director of the process of protein synthesis and RNA carries out the instructions.

Sarcoma: A type of cancer that forms in the connective or supportive tissues of the body such as muscle, bone and fatty tissue. Sarcomas account for less than 1% of cancers.

Septicaemia: Blood poisoning.

Stem cell: A cell that has not yet acquired a special function. In a developing embryo, stem cells may differentiate into all of the specialised embryonic tissues. In adult organisms, stem cells act as a repair system replacing tissues damaged by disease or injury.

Subarachnoid haemorrhage: Bleeding between the brain and one of the covering membranes, often due to a ruptured aneurysm, a weak spot in the wall of a blood vessel.

Subdural haemorrhage: Occurs when a blood vessel ruptures and blood builds up between the brain and the brain's tough outer lining, the dura.

Surgical pathology: The study of tissues removed from living patients during surgery to help diagnose a disease and determine a treatment plan (*see also* **pathology; anatomical pathology**).

Thrombus: A thrombus is a blood clot that forms inside a blood vessel or cavity of the heart.

Trisomy 13: A genetic disorder in which each cell in the body has three copies of chromosome 13 instead of the usual two copies.

Ventricles: The lower two chambers of the heart.

INDEX

history of 158, 159–60
Lauren Ackerman and 147–8,
150, 161, 162
molecular biology and 131–2
scientific pathology and 161
susto (startled) 352–3
Sweeney, John 245–6
syphilis 210

Taubenberger, Jeffery 6, 127–46
teratomas 347, 348
Thailand 48
Timperley, Walter 191
Tissue Bank 270
tissue engineering 11, 20–23, 27
tissue retention 84, 251–2
families and 251–2, 271–2
'presumed consent' on 260–61
value of 294
see also Armed Forces Institute
of Pathology (AFIP); Human
Tissue Act; Tissue Bank
transurethral resection (TUR) 155
trophy skulls 95
Truth and Reconciliation
Commission, South Africa
327–8
tuberculosis (TB) 68, 73, 76, 78,
170, 213, 219–20, 237, 303,
341
in Côte D'Ivoire 80–81
see also AIDS, HIV
Tunisia 4, 168, 174
Turkey 168, 174
ultrasound 40–41, 216
United States Medical Licensing
Examination (USMLE) 37
University College Hospital (UCH),
London 69, 73, 80

valvular disease 219
van den Ende, Professor Jan 234
Van Velzen, Dick 83
see also Alder Hey Hospital
Vanezis, Peter 55–6

Vass, Arpad 91
Vermont, University of 230, 238,
240–41
Verryn, Paul 327
Viagra 11
Virchow, Rudolf 159–60, 300
Virchow's Archive 46
viruses 2, 6, 74, 144
efficiency of 127
endogenous retroviruses 128–9
H5N1 bird flu virus 140, 143
human papillomavirus (HPV)
224, 231, 233
transmission of 77, 78
see also HIV; influenza; Spanish flu
von Hagens, Gunther 8

Wainwright, Helen 5, 204–23
Wainwright, Professor (Natal
Medical School) 229
Waluś, Janusz 323
warfarin 217–18
Warren, Robin 267
Wei Hsueh 5
Wigglesworth, Jonathan 287
Will, Bob 194–5, 197, 202
Wilson, Liz, case of 102
Witwatersrand, University of,
Johannesburg 224, 234, 236,
237–8, 313
women
and guilt 25–6
in pathology 223, 315–16,
317–18
in science 24–5
World Health Organization
(WHO) 114, 202
Wright, Nicholas 298–312

Yacoub, Sir Magdi 14, 15, 17, 19,
25
Yale University 28–9, 37, 46

zur Hausen, Harald 231

375